MW01052827

"PILOT'S DISCRETION"

This book is based primarily upon personal experiences of the author, and the information presented has been found relatively safe and satisfactory in his hands. He cannot, of course, be responsible for any difficulties that the reader may have as a result of using the advice contained herein. Pilots using any of the flying recommendations must be sure to consider them in light of their own experience, proficiency and equipment. Similarly, some of the non-flying suggestions concerning such things as wilderness areas, hypothermia, wild animals, etc., discuss situations with probably much more potential hazard than flying and, again, the reader must measure them against his own skills and experience.

FLYING--
Off the pavement!

By Link Grindle

A manual for the Amateur Bush Pilot

Cartoons by Pegi Wear
Cover by George Carll
Photos & Diagrams by the Author

LASENDA PUBLISHERS
SOUTH LAGUNA, CALIFORNIA

Grateful acknowledgement is made to the following for permission to reprint published materials:
Cartoon *TUMBLEWEEDS* by Tom K. Ryan, copyright ©1976 by King Features Syndicate.
Quotation by David Brower from *BAJA CALIFORNIA AND THE GEOGRAPHY OF HOPE,* by Joseph Weed Krutch and Eliot Porter, copyright ©1967 by the Sierra Club (used with permission of the publisher).

Library of Congress Catalog Card Number 77–76156

ISBN: Soft cover 0–918916–01–1
Hard cover 0–918916–02–X

Published by and copies available from LASENDA PUBLISHERS, 32331 Coast Highway, South Laguna, California 92677.

First printing – May, 1977
Second printing – August, 1977
Third (revised) printing – January, 1978

Dedicated to –
B.W.T.C. & O.

TABLE OF CONTENTS

TABLE OF CONTENTS

FORWORD

THE SEARCH FOR SOLITUDE

The inspiration for this book first came to me while falling asleep in a little tent in a comfortable sleeping bag on a soft foam mattress on the quiet shore of an isolated lake in Northern Idaho. It had been an interesting day. The last day of a professional meeting held at a prestigious luxury hotel in a large colorful seacoast city. The meeting had been excellent; the scientific sessions in the soft-chaired conference rooms were interposed by gourmet meals and with rounds of tennis, golf, boat rides in the harbor and chartered tours of places of interest. It had been the ultimate in refined luxury living; the epitome of a civilized vacation.

And people! People everywhere! Nice people, kind people, old-friend-type people – amid a background of thousands of unknown faces. It had been a relief to say good bye to friends, leave the congested city, climb into the small plane and make a short flight to this region of soft and quiet solitude.

I have a happy, satisfying life with time more than filled between a successful and fulfilling professional life and a loving but busy and active family. But more and more I find the need for periods of solitude, for rest and relaxation without demands from any people, and more and more the usual vacations do not seem to satisfy this need. Most vacation spots are becoming increasingly impersonal and commercial. Airport waiting rooms are crowded, smokey and noisy. Hotels often seem to have some slightly annoying activity going on nearby. Motels always seem to have at least one occupant who slams doors at night or noisily warms up his engine at dawn.

The current popularity of campers and motor homes shows that these feelings are widespread. They indicate the increasing need of more and more people to escape these population pressures. But even these escapes are having increasing limitations. As recreation vehicle popularity spreads, it takes longer and longer to drive far enough to find real solitude. And even then, the tranquility is often tenuous. The next space may suddenly be occupied by a crowd of boisterous beer-bottle-throwing recreationists, perhaps spending much of their time tuning up and road testing noisy racing-type motorcycles.

Another similar manifestation of man's search for solitude is the current popularity of backpacking. A visit to any one of the mushrooming number of "trail shops" gives an indication of the number of people that are today feeling the need to escape from our increasing "civilization," and it is also an indication of a somewhat self-defeating aspect of this trend. It is becoming necessary to pack in further and further from the trailhead to get away from the increasing hoards that are hiking in the backcountry today.

The attraction of the wilderness has been an important trait of man since before the start of recorded history. The human race has always had the urge to explore the unknown. A good part of history is the story of explorers and their findings. The Norsemen, Columbus, Marco Polo, Magellan, Lewis and Clark, Admirals Peary and Byrd are just a few from the long, long list of famous explorers and seekers of unknown lands. Following them thousands of other humans all over the globe and throughout the history of man have had an inherent urge to follow into the wilderness, to find the frontier and to remain and settle in isolated areas.

But now, for the first time in man's history, there is no frontier. There are no major unexplored and undiscovered areas of the world to search for and to settle in. Man's explorations have had to turn to the heavens; and here no one, as yet, can follow. Yet, there remains within many of us, strong remnants of this exploratory heritage. We can no longer explore for new lands, but we can experience some of these same feelings by seeking the remaining unsettled and isolated areas. One of the few escapes left today for those of us with strong Daniel Boone or Thoreau-type instincts in our makeup is exploring the wilderness with a light airplane and perhaps also camping there. There are still in much of the North American Continent airstrips in isolated areas accessible only to the lightplane, airstrips which are completely away from any contact with civilization.

Recently, space age technology has enabled camping equipment makers to create camping gear of amazing light weight and utility which makes very comfortable camping easy and practical with a small airplane.

This book is based on experiences starting over thirty years ago with camping by truck in the then very isolated areas of the American Southwest, which are now overrun by fences, No Trespassing signs and luxury hotels. It concentrates specifically on my experiences and experiments of over ten years with light plane exploring and camping, the best way left today to really get away from it all. Over this period I have slowly evolved techniques that make it relatively easy and simple for me to find such isolated airstrips and to enjoy wilderness flying experiences. Many of my flying friends, a lot of whom are much better and more experienced pilots than I am, have expressed an interest in duplicating some of my adventures as well as indicating a lot of uncertainty about how to do it. I have heard this so frequently that I felt that a book explaining these techniques might be useful for many fliers. I present it humbly and with considerable humility. These are techniques which have helped me and which work for me, but I am fully aware that many of my ideas will be "wrong" for someone else. One of the fascinations of flying is the differences of opinion that exist between pilots; indeed, it sometimes seems that most conversations between pilots are really friendly arguments about various points of flying. I will be happy to receive any comments, suggestions and criticisms and will do my best to respond to them.

There are already a number of excellent books which describe quite exotic places to fly to in many areas of North America. They contain very valuable data about interesting airports and how to get there. This book is not intended to be such a "cookbook" with such specific recipes, but rather is a guide on how to *find your own* isolated places to fly to. There are suggestions as to the different flying proficiencies that may be required to fly to these somewhat different areas. There are also suggestions for choosing proper clothing and camping gear, considering the neglected needs of the lightplane flier rather than those of the backpacker or automobile camper.

The specific guide books are valuable and you should have them, but it is important to be able to find your own locations, because just by being listed in a flying manual, a good spot for isolation often becomes popular and the listing is thus self-defeating.

Even so, as examples of the choosing process, I have included a number of places that I have personally located by the techniques to be described and that are interesting and relaxing. But, the main purpose of this book is to show you how to enjoy finding your own *Shangri-La's.*

A great many people both at home as well as on my journeys have given me much help, information and encouragement. It is impossible to thank them all here, but I am grateful to them all and especially to Orson W. White, M.D., a long time professional and flying friend and a pioneer lightplane explorer, for his multiple suggestions; to my airplane partner and long time Sierra Club trip leader C. Edward Miller for his valued advice about camping and the wilderness, and to his wife Jean Denney Miller for help in showing me the feminine viewpoint about it all; to Marvin Patchen, founder of AERO Magazine, for his encouraging suggestions in the early stages of this book; to Patricia Grignon for her skillful editorial help; to Ellen Peebler for her artistic help and guidance for my "off the pavement" trip through the fascinating unknowns of book printing and publishing; to Sra. Anarosa Pelayo Jones for discussing the Mexican viewpoint with me; and finally to my wife Betty and my sons Wade and Tom for their suggestions and encouragement and their patience with all the time I was away from them, both by airplane and by typewriter. None of these should be held responsible for any errors or omissions that this book may contain, because I did not always accept or follow their advice.

Lincoln Grindle

Bahia San Luis Gonzaga
May 1977

PART I
THE WILDERNESS

CHAPTER 1

INTRODUCTION

WILDERNESS FLYING • LIGHTPLANE CAMPING

This book concentrates on two phases of flying that are not covered in most other aviation publications. The first is locating and using isolated, usually non-paved strips where civilization can be left completely behind and the relaxing atmosphere of the wilderness can be enjoyed. The second is the related subject of camping with an airplane with emphasis on equipment and other requirements which are unique for the lightplane.

WILDERNESS FLYING

To fly into wild areas requires knowledge and techniques which are different in some critical ways from those required for the usual civilized country flying. Navigation is different. Mountains and weather may be unfamiliary to a flier from flatter country. Locating, evaluating, landing on and taking off from isolated non-paved strips requires a consideration of new factors. To the novice to wilderness flying, contemplation of these new things often seems both complicated and frightening.

While a good understanding of these differences is very essential for safe flying in the wilderness, they are actually less complicated than many phases of regular flying. Most of them are, really, facets of regular flying that are just used in a little different perspective than usual. This book hopes to point out many of these differences and while emphasizing their importance, at the same time to make them seem more simple and non-frightening.

LIGHTPLANE CAMPING

Camping with an airplane is not a new concept. It is as old as the airplane itself. The early fliers frequently slept and ate under their wings. In fact, they often had to, because of the lack of accommodations near flying areas in those days. Their overnight resting spot was usually a smooth field or pasture (the origin of the term "airfield"). There might be a friendly farmer nearby who would offer a night's lodging and meals, but more often there was not, and the fliers were on their own. They had to be a hardy lot, because the camping equipment of those days was relatively primitive, rough and so bulky and heavy that enough of it to provide really comfortable conditions was beyond the carrying capacity of those small low-powered early planes. Yet, the old-time pilots would roll in a blanket after eating a can of baked beans and seem to be as good as new in the morning.

Over the years, the "airfields" have evolved into specialized areas for the lightplane and the pilot. Modern and comfortable accommodations are located at, or easily accessible to, the airfields so that now most RON pilots eat in semigourmet restaurants and sleep with the same comfort that they have at home. The need for camping is gone and, in the minds of many pilots, so is any plausible reason for it.

And yet, there are hundreds of pilots who still occasionally sleep with their planes. They do this for reasons of economy, convenience or often they like to go and stay at places where there are no suitable overnight facilities. But their camping gear is often makeshift, incomplete and much less comfortable than it could be, because they have not kept up on the recent advances in modern camping equipment which today can be ultra lightweight and give comfort rivalling that at home.

Much is written about camping and much about flying; but very little about the combination of the two. Camping catalogs and stores are oriented to the backpacker, automobile camper or boating enthusiast. The needs and requirements of the airplane camper are not known or understood by most of the camping equipment industry. They have gear that is wonderfully suited for the airplane, but they do not recognize it as such, because they are not oriented to the airplane or its requirements. The airplane camper has had to interpolate the data about on-the-ground camping equipment for his use in the airplane. Frequently he is very successful, but often he is not; and sometimes he misses items that have excellent application for the plane. One purpose of this book is to evaluate present day camping equipment and methods for the flier's unique needs and uses.

There are, today, a great many pilots who could benefit and expand their use of their planes by doing some camping, pilots who just never consider it, or if they do, reject the idea because of erroneous pessimism about it. This book hopes to give these pilots new perspectives about airplane camping and help them reconsider its potential benefits.

Camping is an essential background for wilderness flying because usually, but not always, camping is necessary for staying in a primitive area since the lack of accommodations is one of the main reasons why it is still primitive. Camping by airplane is not limited to primitive areas, however. Proper techniques for airplane camping can also be very helpful and time saving for the private plane traveller who has no interest in getting out in the wilderness. So the two main areas of this book, locating the wilderness and camping are, in many ways, related, but can also be completely independent.

CHAPTER 2

WILDERNESS IN PERIL

Before the second World War, getting to real wilderness was not simple but it could be done relatively easily. In those days, prior to turnpikes and interstate highways, cross country drives could be veritable expeditions, into country that was wild and unspoiled, unroped and untamed. Many wonders that have long since been fenced, posted and paved to accommodate today's travelling masses were then isolated curiosities often reached only by miserable unpaved roads. There were no signs, no fences, no one to enforce the since-developed regulations. Often no one even to give directions or advice. Often just no one.

But there was a feeling of adventure, of discovery and of exploration. There was a freedom to wander anywhere and everywhere, unregistered and undirected, unregulated and unrestricted. Freedom to walk the river valley, to wander into the ruins, to visit the then uncommercialized Indians – the visitor in those days was as much a curiosity to the Indian as the Indian was to the visitor. This was the ultimate relaxation. The peaceful frontier where there was a feeling of being on the very edge of civilization and of obtaining strength and peace from experiencing the untouched primeval areas just beyond.

This need for experiencing the wilderness is a basic part of most human makeups and has been expressed in many ways for centuries. David Brower, former executive director of the Sierra Club has written, "Without . . .wilderness, the world is just that much closer to becoming a cage." (1)

In those bygone times, it was relatively difficult to reach this wilderness frontier, but it was there and, with effort, it was reachable. The very improvements in transportation that made it so much easier and simpler to reach, at the same time destroyed it by making it so reachable. A fragile thing, it could not stand the influx of all those who now could easily find it and its very popularity obliterated it. The Four Corner area was once a mecca for isolationists; for those few who could locate horses and ride into its barren grandeur. It was spoiled, first by the gradual onslaught of the four wheel drive recreation vehicle, and finally by the paved parking lot adjacent to the new highway, where the litter cans hold only part of the litter left by the hoards who briefly stop there because the sign on the highway says they should. They have slight curiosity and less interest. There is no longer even a trace of that grand isolation of the primeval wilderness. Instead, there is just a pause from the continual ride, another "must see" to be checked off on the travel guide, another amusement park type attraction easily seen and easily forgotten. Any remaining feeling of wilderness or relaxation is now as ersatz as the wilderness areas so artificially created at the amusement park.

For may people, perhaps for most people, the amusement park type experience is very satisfying and it is enough. Perhaps these are the lucky people. Because it is obvious that the real wilderness is nearly gone and that its artifical replacements are about all that is left.

But there are still some people who crave more than this artificiality. They are the emotional descendants of the explorers and the pioneers who still feel the need for and get satisfaction from the *true* wilderness. They try to find it by backpacking, cross country skiing and various expeditions and tours, often sponsored by wilderness oriented organizations. But there is increasing frustration for these seekers, as the true wilderness dwindles away on it from all sides. The very popularity of the wilderness is causing its destruction because it survives only without popularity. However ecology minded, litterless and quiet the new visitors may be, they will, just by sheer numbers, have an unavoidably deleterious effect on the wilderness of the trail. Wilderness equates to solitude, and solitude is long gone in such a multitude. Add to this the more avaricious onslaught of the logger, the developer and the commercial miner and it is obvious – the wilderness cannot long survive.

The notorious Alaska pipeline crosses one of the largest, grandest and most isolated areas left in the world. Formerly, it was only partly known and then only by the backpacker or the float plane pilot. But now it is cracked open. Isolated frontier villages now hear the roar of the deisel. Wolves are chasing pickup trucks for handouts. Bears are begging around work camps for snacks of junk foods. The animals find these pursuits easier than working for their former natural health foods.(2) It is only, so far, a thin scar, but a scar described by Averill Thayer, manager of the Arctic National Wildlife Range as, "Sort of like a razor blade across the Mona Lisa."(3)

In the National Parks even the Ranger-Naturalist unavoidably damages the wilderness, because the necessary steps taken to protect the areas from the crowds as well as to accommodate these increasing crowds, largely remove the freedom of the true wilderness and partially convert the atmosphere into the amusement park unreality. The sign, "Stay on the Trail – Avoid Erosion" is necessary because if only a few of the many thousands who visit there scrambled up the bare hillside, they would cause a nearly complete destruction. Yet, at the same time, the sign shatters the wilderness for the true believer, "Keep Off The Grass" and one is again in the city park and can hear, in fancy, the roar of the traffic on the nearby city streets.

The Ranger-Naturalist is there to enforce the sign just as the cop is near the park with a citation for anyone found on the grass – in both cases very necessary but the very antithesis of the true wilderness spirit. The city park became urbanized because of the increasingly dense population of the city; now the National Park is becoming urbanized as the population increases there.

If there is any hope for the preservation of the remaining fragments of the wilderness it is that some of the public has finally become aroused to the dangers confronting the wilderness and is making protective moves towards it. It is more and more difficult for the logger and the developer to proceed roughshod.

The government is putting more land into preserves of various sorts, and land already in preserves is being treated more considerately with increased resistance toward the exploiter. The tide is being slowed, if not actually turned, against the loss of more wilderness. These are encouraging trends and must be supported by all who love the outdoors.

Still, the overall outlook is pessimistic. These encouraging trends are up against some almost irrestible forces – the increase in world population and the increasing mobility of this population. People have a right to the wilderness, which should not be taken from them, yet their increasingly easy and inexpensive transportation to the wilderness areas is its greatest danger. Those of us who fly lightplanes can use them to give us a greater enjoyment of the remaining true wilderness to a degree that can be done in no other way. We can also enjoy the knowledge that with proper care on our part we can participate in this true old time wilderness without causing it any harm or destruction.

CHAPTER 3

ISOLATION VERSUS LUXURY

The areas of the continent that are still wild enough to be enjoyable for their wilderness are far from the beaten path. They exist in small pockets in the lower 48 states, small Caribbean islands, or else as larger areas such as much of Northern Canada, some of Mexico or most of Alaska. It is usually distance, expressed either in miles or as roughness of any roads to it, that keeps an area in its unspoiled condition. Basically, this is a matter of time, whether the time required to go the necessary distance, or the time needed to slowly traverse a primitive road. And time is where the lightplane makes the difference.

All land was originally entirely wilderness; and today various areas have been subject to various degrees of civilization, so that areas can be found with all degrees of isolation. Thus, there is a graded spectrum of possible outdoor vacations available to the private flier, starting with the very civilized luxury resort which most resembles the big city country club but with the city removed and good scenery replacing it (often viewed only through the picture windors in the lounges). The spectrum blends on up (or down, depending on one's point of view) to less and less luxurious accommodations with more and more of the attributes of the wilderness. Finally at the other extreme there is the complete isolation of the float-plane pilot camped on the edge of an unknown lake.

To each, his own. Everyone can choose the degree of luxury in contrast to the primitive atmosphere he wants. The purist will insist on complete isolation and will be happy to pay for it by living in the rather spartan conditions that this necessitates. But the completely isolated virgin wilderness may, by its very primitive starkness, be too harsh and uncomfortable for real enjoyment by many of those accustomed to the comforts of civilization. These folks will be happier to give up a little of the wilderness and to have in return a little more of the customary comforts. For them, these comforts more than make up for the loss of complete isolation.

The luxuriousness of a resort is often related to the type of air service that it has. Those based near a large paved airport with scheduled airline service will usually be the most deluxe. The rougher and poorer the strip, the more primitive the associated resort will usually be.

It is important to emphasize that a combination wilderness-luxury vacation can often be accomplished with a lightplane. A luxury resort with an airstrip can be chosen in areas where a relatively short flight can take one out for the day to another strip in complete solitude and isolation. This is often a very happy solution for those who really enjoy spending time in true wilderness but enjoy it more if they can sleep and eat in luxury surroundings. Sometimes one spouse will enjoy staying at a luxurious resort while the other will enjoy getting out in the wilds to fish, pan for gold or just goof off. With a lightplane, and frequently only with a lightplane, this is possible and both partners can have different vacations and yet be together. (If this is attempted, be sure that the weather will remain safely flyable for the day and that adequate emergency overnight gear is always in the plane.) There are also some fliers who love the wilderness but may not want to spend the entire vacation there. With a plane, it is possible to stay as long as desired in the isolated area and then fly to a resort with golf, tennis or whatever, for the rest of the time.

PART II
FLYING CONSIDERATIONS FOR THE WILDERNESS

CHAPTER 4

INTRODUCTION

This section assumes that the reader already has had adequate flying training and experience to be a safe flier. It is not intended as a flying instruction manual. Its purpose is rather to emphasize some of the ways that flying in the wilderness *differs* from "ordinary" flying. You will not find in here, for example, descriptions of the actual techniques for short or soft field landings, or descriptions of the actual methods of determining density altitudes. These things, along with all other basic flying techniques, are already very adequately explained in many excellent books on actual flying procedures. Instead the concentration will be on showing *how to tell* when you may need to make a short or soft field landing, and how to *recognize the conditions* that indicate the need for calculation of the density altitude. In these things, as well as in many other facets of wilderness flying, there is a gap in the sources of information between the actual flight training books and the books that describe unusual spots to fly to. This book will attempt to fill in some of this gap and to show how to determine when certain previously-learned techniques may be needed en route and especially to show how modifications of usual flying techniques may be needed in wilderness flying. It is not basically a "How To" book, but rather a "When To" book.

Many of today's pilots are trained on long, wide paved runways and do all their flying from one such runway to another. Actual emergency landing practice and other off pavement maneuvers are nearly abandoned at present. Flying in backcountry areas and using dirt strips requires a number of considerations that are not required for ordinary "pavement" flying. From the standpoints of both safety and comfort, there are a number of different parameters that must be considered. Actually, the difference in using a non-paved strip is largely psychological because of excess concern with possible variables not usually found on a paved strip.

The Proper Instructor

A number of times this chapter will suggest that it might be advisable to enlist the aid of an instructor before first practicing some unfamiliar wilderness flying technique. When you choose an instructor for this, it is extremely important to pick one that not only understands the points that you want to practice but also to get one that is adequately familiar with your particular plane. During his biannual flight check, a friend of mine was severely criticized by the instructor who was examining him for delaying the retraction of the gear on a practice go around. It was in the friend's 210, and on his particular model the gear and gear doors move into position of severe drag as the retracting cycle starts. This is specifically described in the plane's operational manual and the delay is recommmended procedure. Fortunately, the friend was able to point this out to the instructor tactfully enough to pass the check ride. But the point is that this instructor, obviously familiar with other planes, still would not be a good one to practice wilderness techniques with in a 210.

CHAPTER 5

THE OFF-PAVEMENT LANDING

STRIP LENGTH • STRIP SURFACE • COME DOWN CAREFULLY
WIND DIRECTION • WHERE TO PRACTICE • STOL

Most of the significant differences in flying technique for wilderness flying relate to the landing. Indeed, the characteristic most often associated with the legendary "bush pilot" is his ability to land in places others can't (or won't). Major differences in off pavement landings relate to the length of the strip and to its surface.

STRIP LENGTH

The length of a strip means its *safely* useable length, and it is essential to be able to determine from the air how much of a strip is safe to use. It is important to know how much length the plane needs and, more important, to be able to judge this length from the air. Actually, it would be possible, of course, just to always allow a lot of extra length in the estimate to be sure, but this will prevent trying some strips that could safely be used. Some extra margin for safety must always be allowed in estimations, but the more precise the estimations of length from the air, the more strips will be safely available.

Minimum necessary length is not emphasized in routine training, nor does it usually need to be, because most paved strips used in normal flying have far in excess of the needed length. Proficiency in estimating minimal safe length depends on four parameters: learning the actual minimal landing roll of your plane; estimating this length from the air; practicing landing techniques that will shorten the needed roll; recognizing other factors that will change the needed length.

Length Actually Needed

The true needed length can be determined only by actually measuring it. The plane's operational manual will list a minimal landing roll but in reality this is only a ballpark figure. To know exactly, the plane has to be actually landed as short as possible and the spot where it stops noted carefully. (This can be a little hard on brakes and tires, but it is possible to ease up a little and fudge by estimating where it *could* have stopped and be pretty accurate and still have round tires.) Mark this stopping spot either by reference to something abeam the plane on the ground, or actually place a visible object beside the strip at this point.

Estimating Needed Length

Then go up and see how this distance looks from various altitudes. This is the only way to learn to judge the minimal landing distance from the air. Don't be misled by the optical illusion sometimes created by the relative widths of strips. Usually, we look at wide runways with flat cleared areas beside them. A wilderness strip, by contrast, is often narrower and

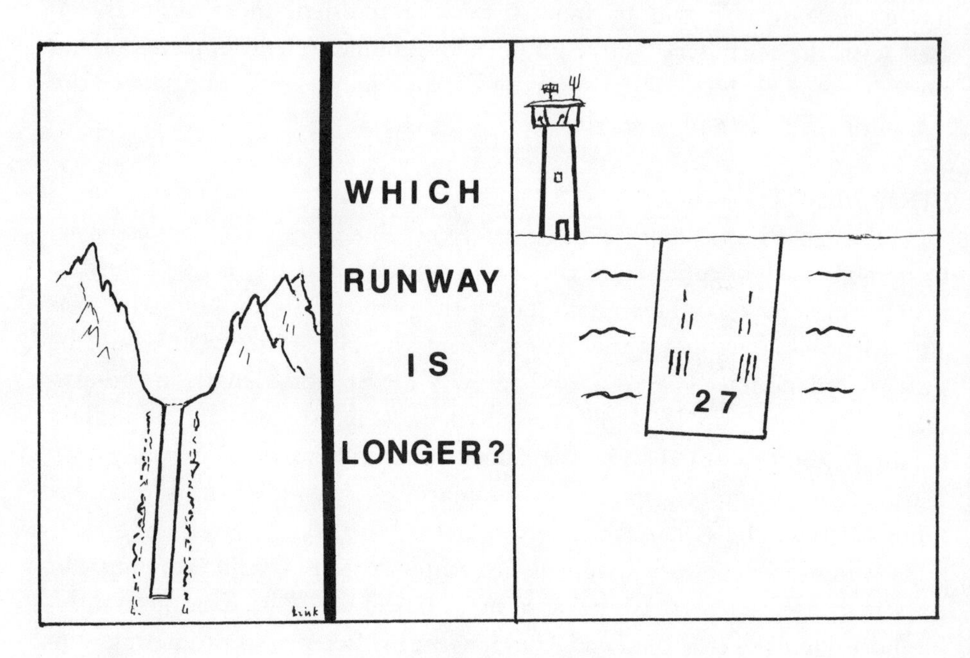

has vegetation right along side of it. This makes it look longer than it is.

Shortening Your Needed Length

To learn how to land in less length, the aid of an instructor may be advisable. Flying low and slow on approach can be dangerous and an instructor who is familiar with your type aircraft may be needed to keep from overdoing it. But careful instruction and lots of practice in precision landings can significantly cut the amount of landing space needed.

Factors Affecting Needed Length

The other factors that will vary the landing roll are more or less common sense principles of physics which are familiar to most pilots. **Wind** can shorten the landing roll considerably (in some "one way" strips, a down-wind landing is necessary so that the wind increases the length needed). **Higher elevation** means faster landing speed and more needed length. A **rough surface** will slow a plane down more quickly than the usual smooth surface. **Low vegetation** will shorten a roll while **water or ice** will lengthen it. **Mud** can go either way. It may be slippery and make a plane skid and go farther or it may be gooey enough to slow it down. Mud (which is often hard to see) may show up suddenly under only one wheel and do sudden strange things to directional control. If thick mud is hit with both wheels at nearly the same time, the plane will just stop more quickly. Then, when the door is opened in the middle of a field – well,

that's not really a flying problem. (The plane will need extra care after landing in mud. This is especially true in mud or sand that is wet from sea water. We fly frequently into a strip in Mexico that is covered several times a month by the ocean.[4] This keeps the surface well packed and firm; but is a problem for aircraft maintenance. In spite of extra careful washing every time we get home the brake drums pit and corrode abnormally fast.) On a rough strip it may be best to ease on the usual braking at first to help keep weight off the nose wheel, which means a longer roll.

All of these things become a *mix of uncertainties,* the net result of which can, at best, only be estimated. All that can be done is to try to keep them all in mind and allow some extra margin for safety.

All this has been intended as a guide to pilots whose training and experience has been only on 5000-plus foot paved runways. Frequently when these pilots attempt landing on a shorter dirt strip it becomes a harrowing experience for both pilot and spectators and much less safe than it could have been with more preparation for short fields. Nothing in the above should be an encouragement to cut the acceptable field length to a dangerous minimum. Be as familiar with the plane's landing needs as possible, but always allow a good extra margin for unexpected unknowns. (Incidentally, the wilderness is no place to go in an unfamiliar plane.)

Always remember that just because a strip is there, doesn't mean that it can accommodate your plane. Many dirt strips have been created and are used only for some light weight STOL type plane and perhaps even then only under certain ideal conditions.

STRIP SURFACE

The surface of a strip is probably the most important consideration in off-pavement flying. A paved strip can usually be assumed to be smooth and flat, especially if it is in regular use. If there are surface irregularities, they usually show up against the even appearance of the rest of it. Water and ice are about the only surface variables that usually occur.

The non-paved strip, especially the infrequently used non-paved strip, is more of an enigma. Whether the surface is dirt, turf or rocky, the irregularity of these surface components, as contrasted to a smooth flat pavement, can mask larger irregularities. The irregular pattern of a rocky or turf surface can blend invisibly into other large dangerous irregularities so that they are sometimes virtually impossible to spot from the air. Dips and ditches from water erosion, spots of soft sand or mud, animal holes, large areas of depression or elevation as well as other anomalies may be completely camouflaged by the overall irregularity of the surface.

If there is longer grass or weeds, all sorts of hazards can be hidden: rocks, ditches, boards – nearly everything that could ruin a landing may be lurking under the soft uniform-looking vegetation. Adjacent hilly terrain may give an optical illusion of perfect flatness, but the strip may actually run significantly uphill or downhill.

How can these things be determined before landing? Often there just is no completely foolproof way, but there are clues which can help. A recent on-the-ground inspection by you or someone you can really trust is the ideal way; but this is usually not practical. If other planes are on the ground, it is an encouraging sign, but not an infallible one. Try to determine the type of planes on the ground. A Super-cub with flotation tires parked safely at the edge of a pasture is not a green light for a heavier tri-gear. Sometimes you can get first-hand reports by radio from a plane on the ground. But most of the wilderness strips to do not have other planes on the ground.

Consider auxiliary warning conditions. Recent rains may have rutted the strip or left soft muddy spots that can look hard from the air. Springtime needs extra caution. Recently thawed ground can be very irregular and soft, and it may be too early in the year for whatever annual maintenance work it may get. Unfamiliar strips that have vegetation hiding the surface should be avoided, unless the touchdown and most of the landing roll can be done in a cleared area.

"Dry" Lakes

In many arid areas particularly in the western parts of the continent former lakes have dried up completely or partially and are shown on the maps as dry lakes. From the air they often look like excellent landing places, but they are notorious for having unpredictable surfaces. Hidden soft spots and ruts are common. An area of water may be covered with dust and dirt and look flat and dry from the air. If you do use one, land near an edge. This part is less likely to be wet or soft than the center, and if there is a problem you may be able to turn out of the lake to drier surrounding ground.

COME DOWN CAREFULLY

If it looks okay from up high, come down and take a closer look. While descending for a better look, continually monitor the surrounding terrain. Valleys and hills can suddenly turn up that are too steep to fly out of. With every significant decrease in altitude, again watch the horizon. It's easy to concentrate only on the strip itself and neglect the surroundings until it is too late. Coming down is easy, but going back up may be another story. Before going any lower it should always be determined that the plane, with its load, in the existing atmospheric conditions will have room to climb safely.

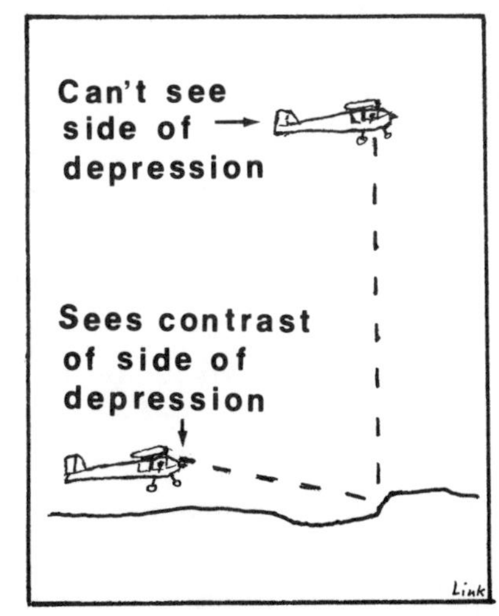

If it is safe to do so, make several very low passes from different directions and inspect the surface. Surface irregularities show up better at very low altitudes because of the flatter angle of view. The lower and slower that flying can be done with safety, the better the surface can be evaluated. Some pilots claim that they can do an instantaneous touch and go and "test the surface" with their wheels. Well. . .may be so.

Early or late in the day, when the sun is coming at a more acute angle with the earth, many irregularities on strip surfaces will show up

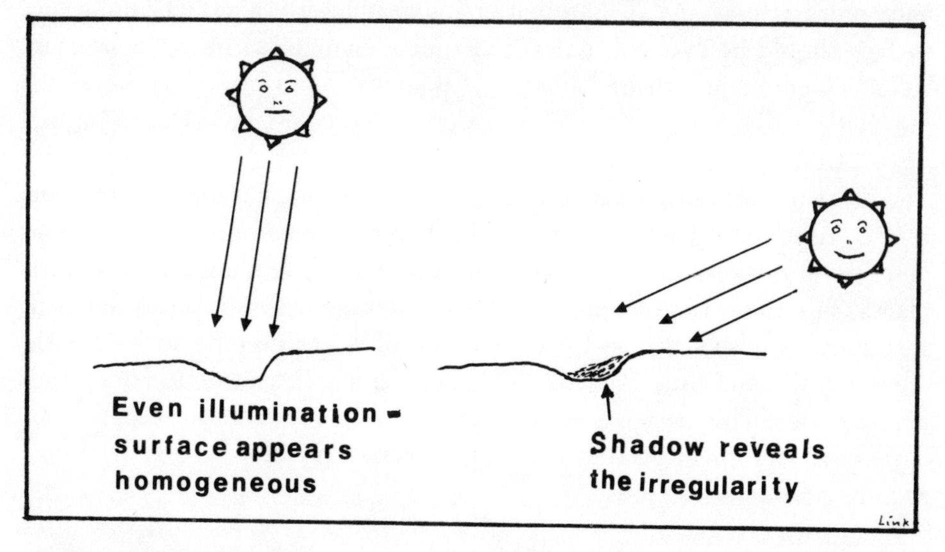

easily, only to disappear completely when the sun is higher overhead. The advantage of the low-in-the-sky sun may be more than offset by having it directly in the eyes on landing or take off.

The look-over process may thus be a gradual and time consuming affair, as the adjacent terrain is gradually slowly surveyed and the air tested for turbulence, downdrafts and other invisible hazards. Even if landing is safe, can a take off be made? Most strips that have been investigated this far will usually be good for at least a "walk-away" landing, but it makes for a happier trip if the plane can leave the strip too!

WIND DIRECTION

The direction of the wind is extra important at marginal strips so that the surface can be touched with minimal speed. At the same time, in the wilderness, it may be difficult to determine which way the wind is blowing. Civilized country fliers are used to getting exact wind reports by radio from the tower or unicom. If these are lacking, there is almost always a windsock or two if not also a tetrahedron. If there is much traffic a plane can often be seen taking off, presumably moving upwind.

Usual Indicators Missing

At the wilderness strip, all this is often missing. No radio help. No traffic to watch. Often no windsock or only a dilapidated one that is hard to see. Or worse yet, one that is rusted into a misleading position. The wilderness usually does not have other helpful features such as cars on dusty roads, smokey factories, wash blowing on a clothesline, flags flapping in the breeze. Some former military fliers carry small smoke flares, but these are difficult for the ordinary civilian pilot and plane. Surface wind direction seems to be almost a sixth sense for many experienced wilderness fliers, who always seem to know it almost instinctively. Actually, they just know the importance of wind direction for a potential sudden emergency landing in the rough, and they have trained themselves to continually watch for clues that are usually missed. These clues can be learned and watched for, but they are only clues and should be considered with caution. Birds usually land, sit or stand facing the wind. So do grazing animals for some reason. (Some I have met on the ground make me suspect that they like to keep their noses upwind from the rest of themselves.)

Wind Patterns On Water

Bodies of water can help. Lakes, large rivers and the ocean show wind patterns that are revealing. Floatplane pilots in bush country become amazingly proficient at "reading the lake," by learning to observe wind patterns.

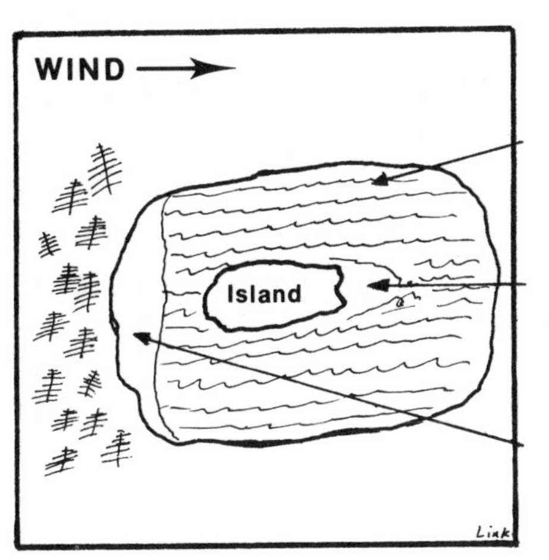

Wind streaks across the water will often narrow the possible wind directions down to two.

Wind patterns will often show a calm area in the lee of an island.

A rim of flat calm can often be seen from the air on the windward edge of a wooded lake.

Boats at anchor always head into the wind; but be careful for boats anchored (invisibly from the air) at both ends. I once got fooled by a boat under way moving imperceptably downwind.

Wind Drift

If all else fails, get down to a safely low level and fly slowly across the strip at a 90 degree angle from its heading and watch the gound for any sideways wind drift. Usually any significant amount of surface wind will be blowing in the same direction as the wind 100 feet above the surface. It's crude and not foolproof, but it's a "straw in the wind," if nothing else helped.

Cloud Shadows

One other last resort is to watch the direction of cloud shadows. If there are broken cumulus or even holes in stratus layers, sometimes you can circle slowly and watch the edges of the shadows on some specific spot on the ground and see in which direction the clouds are moving. Unless there are unusual conditions such as erratic drafts from adjacent hills, the wind will usually be more or less in the same direction on the ground.

Downwind Approaches

Part of the homework for wilderness flying should consist of making downwind approaches, so that the pilot can learn to judge the feeling of the plane's reaction to wrong way wind. This may help the recognition of times when the wind has been judged wrongly. It also helps give a feeling for the occasional downwind landing that is necessary at one way strips. Naturally, this practice will have to be done at a little used strip or some other precautions taken against conflicting with normal upwind traffic.

WHERE TO PRACTICE

Where to practice? If you are not used to dirt strips and smaller strips, how are you going to get the necessary experience safely? First of all, don't go out and locate some nearby minimum sized strip and practice on that. When I first became interested in flying to Mexico, a pilot friend suggested that we go to a nearby strip near San Diego, called Lake Wohlford. This 1500-foot strip is noted not only for its short length, but also for the fact that the sides and one end have abrupt ravines that run down to a lake. All this is made more entertaining by surrounding steep hills as well as squirrelly winds. Another, and wiser, friend said, "Hold on! You can practice and test your short field proficiency just as efficiently and much, much more safely by marking off a Lake Wohlford sized area on a larger airport and using that. Thus a failure to land in this sized strip will not end up as a disaster." This was good advice – so do your practicing on smaller sections of larger strips. When you go for your first "dress rehearsals" on smaller strips, start with less hairy ones and gradually build up your proficiency and confidence.

On a recent trip with a group of doctor missionaries who fly to short dirt airstrips in small villages in the Sierra Madre Mountains in Sinaloa, Mexico, I heard two pilots, chatting on 122.9 ask each other how their short field techniques were developing at these short dirt strips. Apparently it was satisfactory because they were all still flying. But the point is, this is not the place to learn short and soft field technique. It should be learned on small mentally marked off sections of larger strips in the States and practiced there until it is beyond doubt. Dirt strips in the wilds can give excellent practice of short field procedures and build confidence in them, but it is not the place to *learn* these things. (This is doubly true in Mexico where, in addition to the usual physical hazards of small strips anywhere, there is an additional political hazard. Any small run off the strip or other minor accident will frequently land the pilot, if not also the passengers, in jail.)

Crosswind Practice

Proficiency in crosswind landings is even more important in the wilderness than at home. The strips are narrower, as well as shorter, so there is much less room for sideways maneuvering. Also, there is usually only one runway, so a crosswind landing is more likely than at larger places where there is a choice of runways. Again, practice crosswind landings every chance that you can on the larger wide strips so that you will have better control when you get to a narrow one.

STOL

Obviously there are many situations in wilderness flying where a STOL conversion would be very valuable. It could mean shorter and slower runs on rough surfaces for both landing and takeoff. In an emergency landing in the rough, the ability to keep control at slower speeds can allow a ground contact at lower speeds with much less potential danger to the plane and its occupants. Again, its WV^2 and the safety increases with the *square* of the decrease in speed.

CHAPTER 6

THE OFF-PAVEMENT TAKEOFF

IS IT LONG ENOUGH • DENSITY ALTITUDE • LEANING FOR TAKEOFF • TEST IT LIGHTLY • PROTECTING THE PLANE • CHECK THAT GAS

IS IT LONG ENOUGH?

On a wilderness trip out in the boondocks, suddenly the takeoff area seems awfully short. How to proceed? Well, the landing there went safely, so it must be somewhere near long enough for a takeoff--right? Wrong! Satisfactory landing conditions may be entirely inadequate for takeoff. A rough or soft surface works doubly against a takeoff as compared to a landing. It both shortens the needed landing roll and increases the needed takeoff roll. The same can happen at a oneway strip when it is necessary to takeoff downwind or in the rare situation where both landing and takeoff have to be done uphill.

Conditions may have changed since the landing. The temperature may be higher, the wind may be less, some heavy souvenirs may have been taken aboard, a rainshower may have softened the surface and moistened the air. So, calculating the takeoff needs on the basis of what the landing took, as can usually be done on a flat paved strip, is not automatically satisfactory on a rough strip.

Takeoff Length Is More Important Than Landing Length

The importance of length requirements for landing has already been stressed, but the takeoff requirements are even more important. Underestimated length on landing may cause an overrun, but it comes at a relatively slow speed toward the end of the landing roll. Various degrees of bent metal and egos, perhaps, but probably no significant injuries. But, running out of space on takeoff occurs at near flying speeds with therefore much more serious possibilities.

There is also much less excuse for error on take off length. Landing length often has to be estimated from the air, while takeoff length can always be accurately measured. Often there are also various pressures to land which should not be present for takeoff.

Actually, if we are really honest, how many of us really know how much length we need for takeoff? We may have memorized the operational manual figure, but even if we remember it, how accurately can we estimate this as we look from the cockpit down the strip at the starting point? We are all badly spoiled by almost always having much more length than we need. Very rarely do we have a situation where we even get close to our minimum takeoff length. If there is really a question, we can often find the actual length from the radio or a map, although if this aid is available it is almost sure to be a place with excess length. At a strange runway or an intersectional takeoff, we usually decide that it looks "at least as long" as some short strip we know and let it rip.

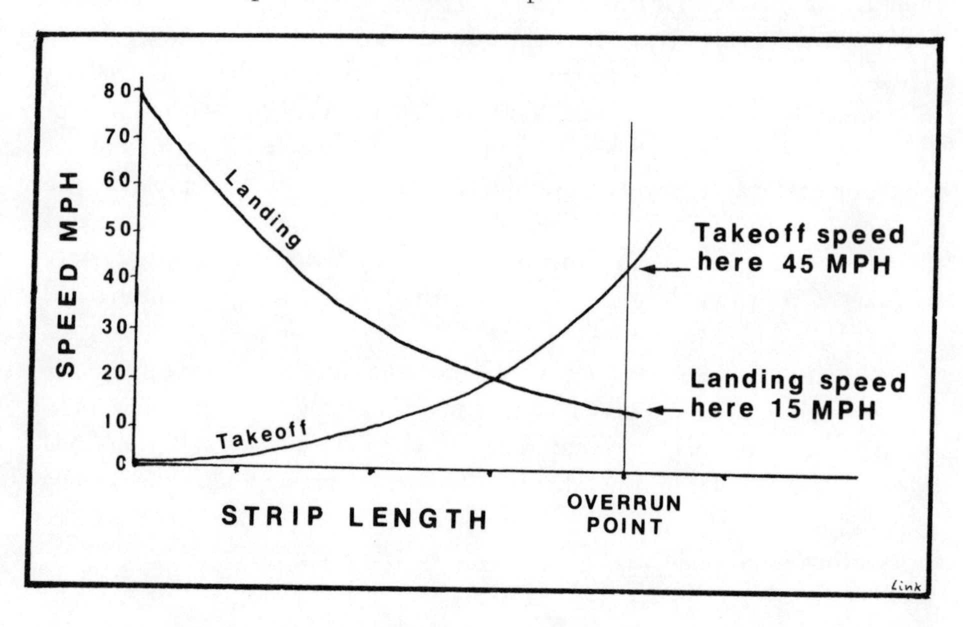

When using isolated strips in the wilderness, every trace of any such complacency must be completely discarded. Very often the extra length we have become used to is not there. Frequently, these strips are hard to build and difficult to maintain and are therefore made as short as possible. Larger commercial planes don't usually use them so the extra length they need is not built in. Often such strips are laid out and built by bush type pilots who are used to getting maximum performance out of their efficient light planes. So it's an entirely different ball game, and there must be an exact knowledge of how much length is needed and how much is really there.

Again, for safety, there must be some homework done beforehand. At a long field at home, someone must watch and determine exactly where the plane breaks ground on takeoff under known conditions of atmosphere and load. In other words, there must be specific measured figures of length needed for gross weight and for some other definitely known lesser amount of load. Use the tables to convert these two figures to "standard conditions," which will then be a base line to calculate actual length needed for wilderness.

Another item of homework is to measure your pacing so that this can be done quickly and accurately in the field. If there is any question about length, pace it off. In the wilds, surroundings of a strip can give optical illusions that make visual estimates extra inaccurate.

DENSITY ALTITUDE

Density altitude is so important that it has been emphasized to the point of making it seem much more complicated than it is. It refers basically to the effect of high temperature and high altitude and especially the combination of the two in greatly decreasing the effective power of a plane and therefore increasing the ground run needed.

If you are not familiar with it, you will find it really unbelievable how much a plane's efficiency is reduced by these factors. Be sure to find it out the easier way; get into the habit of using the tables if there is the least trace of excess heat and altitude. Even very experienced pilots are often surprised by its effect, and it is high on the list of causes of serious flying accidents.

Observe the terrain at the end of the strip. If it is nearly flat – good; if it is a downslope – better; and if it goes up – watch out! The upslope may be gradual and may be masked by surrounding geologic or vegetative features, so look for it carefully.

If density altitude is slowing your climb more than you had expected or if an undetected upslope seems to be coming up toward you, keep the nose down and the speed up as much as possible. All natural survival instincts urge a pull up to get away from that rough stuff, but instinct is completely wrong in this case. It's much safer to clear terrain by fifteen feet at flying speed, than to be fifty feet over it at stalling speed. (Although you may have to explain this later to non-flying passengers!)

In being sure that you have flying speed, don't overdo it. This situation calls for the best angle of climb speed, which for most planes is relatively slow. Retracting the flaps is best done in many planes at a lower speed; there is less sudden sink. But again, it depends on your particular plane. Retractables vary too. Usually getting the gear up just as soon as you are sure you are airborne will help, but in some planes the configuration of the gears and gear doors while cycling can act as speed brakes.

LEANING FOR TAKE OFF

Don't forget to lean the engine as needed for high altitude take offs. This will help give maximum power output and decrease possible spark plug fouling. If you don't know how much to lean, run the engine up to somewhat below take off power and quickly lean until power drop just starts, or if you have an EGT gauge lean to maximum exhaust temperature. Half way between this and full rich is a good estimate. If there is dust or loose rocks as described below, it may be best to omit the run up and just estimate this setting. Most turbo planes will probably do better, of course, using sea level settings.

TEST IT LIGHTLY

Occasionally, the situation will come up where the take off area is critically short and it will be necessary to do all the things that were taught in ground school: Wait hopefully for a stronger head wind, or a cooler time of day, or drier air. If the situation is near borderline, it may be wise to take off alone with no baggage and see how it goes. Often this indicates whether there is enough runway for a full load. Sometimes the only thing to do is take a partial load to some longer (or lower) strip and come back for the rest.

Always keep the plane as light as possible. A few pounds that wouldn't even be noticed on a long paved strip can make a tremendous difference at a short rough strip. The advantage of carrying less gas and thus having less weight must be balanced against the safety of extra gas for the unexpected head wind, bad weather conditions or getting lost.

A usual recommendation for a short field take off has been to put in full power with the brakes on and then suddenly let the brakes go. Thus full power is used for every foot of the runway, which theoretically at least, should shorten the takeoff. Lately this procedure has been under question. Some say that full power cannot be utilized until the plane is rolling well. Others say it is hard on the brakes, but if the brakes are held without any movement against them, it is hard to see how there could be any wear. Even if there is, a little brake wear is a small price to pay for a safer takeoff. For sure, it certainly is one of the worst maneuvers for the plane if dust or rocks are present. As with so many other controversial points in flying, it should be tried both ways (on a safely long strip) and whichever way seems best should be used.

In a critical takeoff situation, some old timers will keep the *cowl flaps* (if any) *closed* until the plane is well airborne. They insist that this much less drag shortens their take off run. If this is done, don't forget to open the flaps up as soon as possible. It's not a usual procedure for after takeoff and it is easy to forget it.

And again, while all these factors are extremely important, you can't learn how to handle a plane by reading about it. If your flying experience has been primarily on long paved strips, get out and practice short and rough field takeoffs, with an instructor at first if necessary. Learn the techniques that enable you to get your plane off the ground best for a short field and on a rough or soft field and on both. Start practicing at your familiar airport and later find a long wide grass or turf field and practice and practice on that. If you can find a large soft field it can give you excellent practice in how different flap settings and other things can affect your takeoff distance. With a trigear, how much to raise the nose is a good thing to practice. If the field is rough, putting the nose up takes weight off the nosewheel as it hits the bumps, but this attitude may keep you on the ground enough longer to be actually harder on it. On a soft field, if the nosewheel can be lifted enough to clear the surface it naturally decreases the drag very significantly, but you will have to see if this advantage will be lost because of the poorer angle of attack that this much nose elevation will cause.

Often pilots lift the nose a little on a soft field takeoff, thinking that they are lifting the nosewheel out of the mud, when all they are actually doing is making the nose wheel strut extend while the nosewheel is still dragging. At the same time their nose high attitude is enough to significantly decrease their lift. They would have been much better off to let the nose alone since the nosewheel is dragging anyway, and get more efficiency from the normal wing angle.

A long high-elevation field can be invaluable for practice, especially if you will be there long enough to try it at different air temperatures. Start practice early at lower temperatures and see how a later increase in temperature effects the takeoff distance. Remember that in spite of the higher ground speeds at higher elevations you always use the *usual indicated* airspeed for both takeoff and landing because the airspeed gauge is effected the same way that the wing and the propeller are.

PROTECTING THE PLANE

There are three extra hazards that are peculiar to unpaved strips, which are important, not so much for safety, but for the protection of the plane. One is bad for the engine, one can damage the fuselage and another endangers the landing gear (particularly the nose wheel).

Dust

The hazard to the engine is dust. Dust, next to excessive heat, is the engine's worst enemy. Unfortunately, many of the wilderness strips do have excessive dust, sometimes so much so that some of it can get past the engine's filters and other protective devices. This is especially true of the very fine silt which has been deposited in many old glacial areas. The particles are fine enough to pass an ordinary filter and, in addition, are extremely abrasive.

Avoid Dust

The best protection against all dust is avoidance. Often it cannot be gotten away from completely, but the engines exposure can be minimized in a number of ways. When taxiing slowly in a light down wind condition (which is the usual situation), the dust blown up by the propeller will often continually flow in the wind past the nose of the engine, or sometimes just keep with it, so that it gets a severe dust bath all the way down to the take off point. Be alert for this situation because it is not always too apparent from the cockpit. Taxiing faster than normal may leave the dust behind; any excess wear to gear and tires by the faster taxiing on the irregular surface will be considerably less damaging than the dust that is thus avoided. If the strip has a wide enough taxiing area, a zig zag course down the field may keep the plane out of the dust for at least part of the taxiing time.

Start Slowly

When starting the takeoff roll, and if the strip is long enough for the extra maneuvering, start out with the lowest RPM's that will start the plane moving. As it gains speed, slowly increase the RPM's a little and then eventually gradually put on takeoff RPM's. By then it should be moving fast enough that the larger cloud of dust made by the takeoff power will stay behind the plane. Getting the nose up as soon as possible and away from the dust may help, but if this slows down the lift off very much the engine may be in the dust enough longer to more than offset the benefit of the greater clearance from it.

If the plane needs a warm up before takeoff, do it in the least dusty spot that can be found and with the plane headed into the wind for sure. Many pilots avoid a run up on dusty strips if the plane has been flown recently and has been operating properly. Mag checks, if really necessary, can be done during the takeoff roll thereby avoiding this additional operation while sitting stationary in the dust. Never check carb heat on the ground in dusty conditions or put it on for at least a few minutes after leaving so that there is time for all the dust to be blown away from all the corners of the air intake. Putting on carb heat makes the incoming air bypass the filters and will greatly increase the amount of dust that gets to the vital areas of the engine.

Often, in spite of all precautions, an excessive amount of dust just can't be avoided. The only remedy then is to change the oil and clean the air filter as soon as possible. It's a drag, but oil is cheaper than metal and the extra change may prolong the engine life greatly.

Always remember that, unlike many other hazards to a plane, dust shows no immediate effect, and it would appear that the extra dust has been gotten away with. But the results will show up much later in reduced engine life. So, avoid it as much as possible and also change the oil and clean the air filter as soon as possible after contact with excessive dust.

This also applies to the rather thick clouds of dust that may be flown through even at high altitudes when there are strong winds on the desert. It does not take very much time for an excessive amount of this fine dust to pass through the air filter.

Loose Rocks

The second hazard is loose rocks which are frequently found on dirt strips. These are easily picked up by the prop, sometimes knicking it severely and sometimes hitting and denting the leading edges of low wings, struts and the horizontal stabilizer. It is essential to avoid loose rocks if possible. If there are safe take off areas without rocks, use them, especially for the first part of the takeoff roll.

If it is not possible to avoid areas of loose rock, at least remove them from the immediate starting area and for some feet ahead of this (unless the rocks are just too numerous to remove). It is unbelievable what large and heavy rocks a prop can sometimes pick up, so try to remove all those that are up to the size of a small fist.

Some of the things suggested for protecting the plane from runway dust will also help it against rocks. Start with slow RPM's at first and get the nose up; but, at best, the prop will take some beating. Planes that fly in Mexico or Alaska can be told by the irregularities of the leading edges of their props. If there are large and rough-edged nicks, they can decrease the prop's efficiency and also cause excessive vibration. If the nicks have ragged edges, the soft metal sometimes allows the defects to "grow" under the constant pressure of the wind. In extreme cases these can eventually end up as a crack through the whole propeller and a loss of part of one blade can occur. This, of course, gives an instant severe out of balance situation which, if the engine isn't shut down fast enough, can loosen the motor mounts – one of the worst conditions you can have in flight. These are extreme cases that could occur only with very severe neglect, but they do show the advisability of inspecting the prop with each pre-flight check and filing down any rough nicks. Most experienced wilderness pilots carry a small file and dress the nicks as they are found. It takes only a few minutes (if a little file is kept in an easily accessible place) and is a worthwhile precaution.

Obstacles

The third hazard is threatening to the landing gear, (particularly the nose wheel) and consists of sudden irregularities that may exist in the take off surface, either small boulders or other objects and also holes, gullies, soft spots or mounds. Frequently these are treacherously hidden by grass or bushes. Unlike the other two hazards, dust and small loose rocks, which were not important at landing, this hazard was also present at the landing. However, it often seems as if these obstacles are somehow avoided on landing, only to show up and become a problem when taxiing or taking off. The "active" part of the strip is often relatively well cleared; it is the taxiing and parking areas that are more likely to have a problem. So before taxiing off to park and before taxiing out to takeoff, unless all the surface can be seen very clearly from the cockpit, stop the engine, get out and walk around and inspect the area for things that could be a problem. Especially when the surface is hidden by tall grass or bushes, be sure and take a good look. Occasionally there will be items hidden that can do damage even to the prop, especially in a tri-gear with its lesser prop clearance. Even if not damaging, a hole can catch a wheel and make it impossible to move the plane without several husky people to pull and tug (and they may not be available in the wilds).

When taking off avoid the obstacles that were previously found. Try to roll over holes and gullies as gently as possible. If the plane is near flying speed it can often "jump" over bad spots or at least lift itself partially and relieve the weight on the gear as it passes the bad spot, then continue the roll normally beyond. Flaps may help increase this lift.

CHECK THAT GAS

It might seem that if the plane has been tied down at a very isolated strip with adequate fuel in it, the usual pre-flight fuel check could be relaxed. Unfortunately this is not the case. There are documented cases where at very isolated places someone has come along and made a "midnight requisition" without leaving a receipt, with disastrous results for the unsuspecting pilot who took off the next morning assuming that he still had all the fuel he arrived with.

CHAPTER 7

FLATLANDER IN THE MOUNTAINS

OFFICIAL MOUNTAINOUS AREAS • EVALUATING MOUNTAIN EMERGENCY LANDING SPOTS • KEEP IT UP • IN A CANYON MOUNTAIN WEATHER • AVOID MOUNTAINS IF POSSIBLE ARE AIRWAYS SAFER? • WHAT'S IN AN HOUR

Flying at cruise in the wilderness, as opposed to taking off and landing, is usually the same as anywhere else. After all, the plane doesn't know any difference between flying over the frequently airported "Golden Triangle" of the northeast United States or over the airportless tundra of Alaska, and basic flying techniques are the same in both places. An exception occurs when mountains are encountered. Because I learned to fly in the varied terrain of Southern California, mountains were just another routine part of flying to me. It was not until I took a trip back to the Cessna factory in Kansas, that I really appreciated and understood what the fuss about mountains was all about. It seemed as if nearly the whole countryside there was one flat safe landing spot and, for the first time, I appreciated how ominous mountains must seem to one who has flown only in these flat areas.

OFFICIAL MOUNTAINOUS AREAS

The FAA designates two official mountainous areas in the United States (excluding Alaska). One is an irregular trapezoid averaging some hundred miles wide running down the Appalachian Mountains from Maine to Alabama. The other is a block covering most of the eleven western states, except California's central valley. This is a great oversimplification, of course, because there are many places with hairy mountains outside of these areas, and within these areas there are many large areas of flatness.

There are courses on mountain flying and many things to read about it. The state aviation departments of several mountain states have published booklets and maps. (They are especially interested in keeping you safe in their mountains so that they won't have to come looking for you.) Some Flight Service Stations in mountainous areas put out good pamphlets with special pointers about peculiar conditions in their area. Newcomers to mountain flying should get and study some of these aids. (5)

Without meaning to put down these courses and publications which do offer much valuable information that may be needed, it does seem that many of them are so anxious to cover all eventualities that they get lost in the process and make mountain flying seem much more formidable and complex than it really is. Actually, if a few relatively simple points are kept in mind, safe mountain flying becomes better understood and easier.

EVALUATING MOUNTAIN EMERGENCY LANDING SPOTS

The first point of difference for the flatland flier is the infrequency of adequate safe emergency landing spots. This will vary with the severity of the mountains. At first, any mountainous country seems, to the novice, to be completely devoid of any safe landing spot. More observation and more experience soon show that there are many more possibilities than first appeared. Nearly all mountains run in longitudinal strips with valleys between them, and nearly all of these valleys contain potential emergency resting places. Not a paved airstrip by any means, but a straight stretch of road, a cultivated field or a sandy bar on a river. These are far from ideal, to be sure, and hopefully will not be needed. But many such spots exist and can be used in an emergency with a little luck without injury to occupants and hopefully with minimal wrinkling of the fuselage.

Much of your increased experience in mountain flying will show up as better identification of these relatively flat areas. As you fly in the mountains, practice evaluating each possibility as it appears. Does a road have a straight stretch that is long enough to land on. How rough and irregular will its surface be, what is along the sides of the road – do trees, poles, wires and fences give adequate clearance for your wings? Roads on sides of hills may be too close to a high embankment on one side for wing clearance, or a drop off on the other side may eliminate it. Look at sand bars along the river beds, and try to estimate their length, flatness and surface. Much that was said about the difficulty in evaluating the surface of dirt strips applies even more strongly to sand bars. Dips, holes and even large rocks can be camouflaged only too well by nature. Still, a river is a good leveler and often a good compressor. If you can avoid large rocks and washed down debris, a sand bar can sometimes be your best bet.

A cultivated field may look inviting, but how soft is the surface; which way do the furrows, if any, go; and what is their relation to the wind? A newly mowed field is probably the best. A plowed field is only a last resort – flip overs are common from the soft furrows. Avoid fields with growing crops, if possible, not only to prevent unhappy farmers, but also because a flip over is very possible if the crops are a certain height and density, and this cannot be determined from the air.

Sometime when you are flying alone, get down low over some of these things and see how your higher altitude evaluation compares with what you find on a close look. This is really the essence of learning about such things, and it cannot be done in a ground school classroom or by reading about it. It is safer to do this low-looking check on your higher impression at areas where a safe-for-sure landing spot is reachable. (You could, for example, check out a river sand bar beside a wide deserted highway that you could use in the very rare occasion that you might need it.) In this way you can more safely get down quite low and really see what the ground situation is. Always observe the precautions that were discussed in evaluating a strip to be sure that you can get safely back up again.

I always used to believe that almost any field that was large and flat enough for a farmer to cultivate with a tractor would accept a light single plane in an emergency. Since I have been looking at more and more such spots from down low, I realize that this is not so. Many cultivated fields that appear level and flat from up high become unbelievably sloped and irregular when seen down low. Farmers must have four wheel drive tractors!

KEEP IT UP!

But do your regular cruising at altitude. One of your greatest friends in the mountains is altitude. The higher you are, the more choices you have to put down in, if you should have to. Often, good altitude will let you cross mountains of considerable height while remaining within gliding range of good safe spots to land in adjacent valleys. Height will also often minimize turbulence as well as enchance communications ability. As you get higher you can see more area and pilotage is easier. Mountain peaks are often essential for pilotage because the section lines, often such a wonderful crutch in flatter country, are usually gone in the mountains. Actually, peaks that appear on the sectional map as if they would be very prominent often do not stand out very much from the surrounding mountains when you are looking at them from the air. At higher altitudes, different mountain peaks are even harder to tell apart. Very clear air can cause the same problem. Distant mountains may be confused with nearer ones; they seem so close. Another old standby, the horizon, is often altered in the mountains. Land and sky meet at various distances above the normal horizon, and the artificial horizon may be needed.

When a ridge of mountains appears directly ahead of the plane, it often appears to be considerably higher than the plane is. A check with the chart may show that the plane is a thousand feet higher than the given elevation for the ridge, but still the ridge looks higher; and it is not until it is almost reached that the optical illusion dissolves and the proper relative altitude becomes apparent. This same phenomenon is sometimes observed when estimating the elevation of an approaching cloud layer.

When first flying in mountains, remember that it's the mountains in front that are important. Don't worry about ones to the side. New fliers (and passengers) to the mountains often feel some panic by the "mountains all around," when actually the flight is up a valley between mountains. There may be no mountains directly ahead but the mountain ranges extending ahead on both sides may give this sensation.

IN A CANYON

When flying in a canyon or up a valley between mountain ridges, the natural place to fly often seems like the center which is furthest from the mountains on each side. Actually, it is safer to fly on one side of the valley or canyon quite near the mountains. The sides are often less turbulent than the center, and if the correct side is chosen (the downwind side), an updraft is often obtained which is of value in several ways. It lifts the

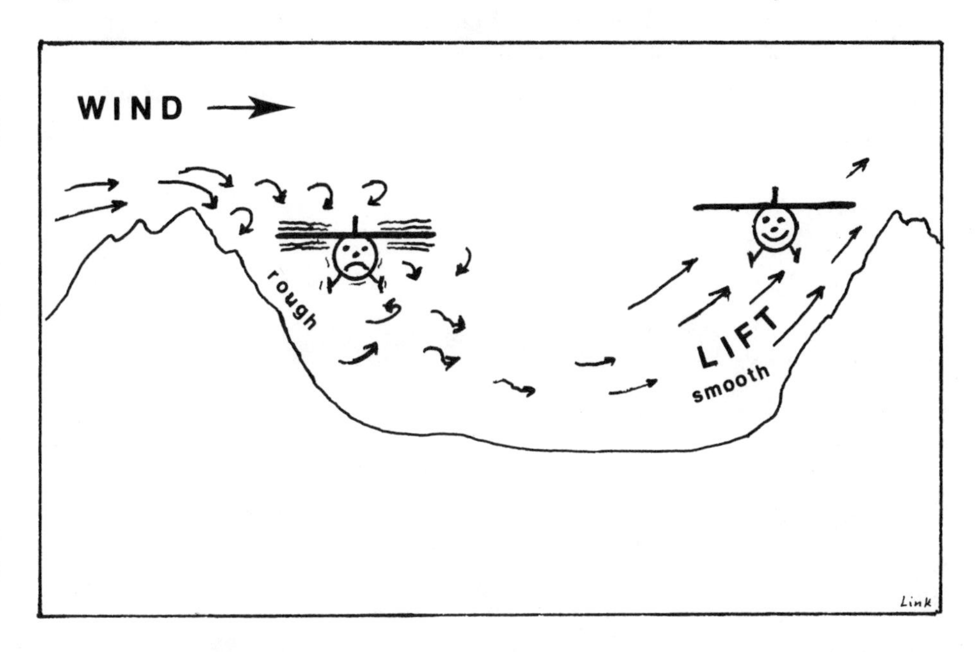

plane and saves the gas that this lift would otherwise require. It helps in getting up above critical elevations faster, or if the plane is nosed down to keep altitude unchanged, the plane will go faster. All the way around, it's a benefit. But most important, being on one side also gives more room to turn around in, if this becomes necessary.

Box Canyons

When flying up a valley with mountains higher than the plane on both sides, always be sure that there is enough room to turn around in. This usually means flying close to the mountains on one side or the other rather than up the middle which would seem instinctively like a safer place. If the ground ahead gives even the slightest suggestion that it may be going to get higher faster than the plane can climb, turn around immediately while there is still room. Either go back or else circle until the plane is high enough to clear safely the higher ground ahead. If the width is already getting too narrow, don't attempt to turn around at cruise speed. Slow to just above stall speed so you can turn as tightly as possible. Slow it by pulling up. This slows it faster than by reducing power and at the same time gets you higher where there is probably more turning room. Full flaps usually give a shorter turn, but you will have to check this out for your particular plane – ahead of time! Every year very experienced mountain fliers are lost because they continue to fly into a blind canyon until there is not enough space to climb over the rising ground ahead and not enough width to turn around in. This must be a very frustrating last feeling, and it can easily be avoided by constantly planning ahead.

MOUNTAIN WEATHER

It should be obvious that weather can be extra important in the mountains. The extra altitude that is necessary for safety in the moun-

Taken near Haines Junction, Yukon Territory *looking across to the Saint Elias Mountains. Ceilings like this at 5000' to 9000', often with rain showers, are quite common in summer in the north country and are* usually *quite stable and unlikely to have sudden changes. Note sunshine on ground in the distance, an indication of* probably *better conditions ahead.*

tains naturally needs higher ceilings. Weather can cause a *reverse box canyon effect* by having a ceiling become lower as a plane flies up a canyon until there is not enough width to turn around in. Continuing is a very ticklish situation because if the mountains at the side cannot be seen, any deviation from the course up the valley will take the plane into the sides of the mountain (and the valley may deviate even if the plane does not). Outclimbing the mountains in the clouds is dangerous, even if VFR on top is reached, because if it should become necessary to go back down again, the miscellaneous unseen mountains will be filling some of the clouds with rocks.

Pre-trip practice should include checking yourself out in slowing the plane, putting on the flaps and doing both, all without gaining altitude. If possible it is good to do this just under a sharp ceiling so that you can experience the effect of going up into this ceiling. Legally, this would have to be done IFR because it would violate VFR cloud clearance specifications.

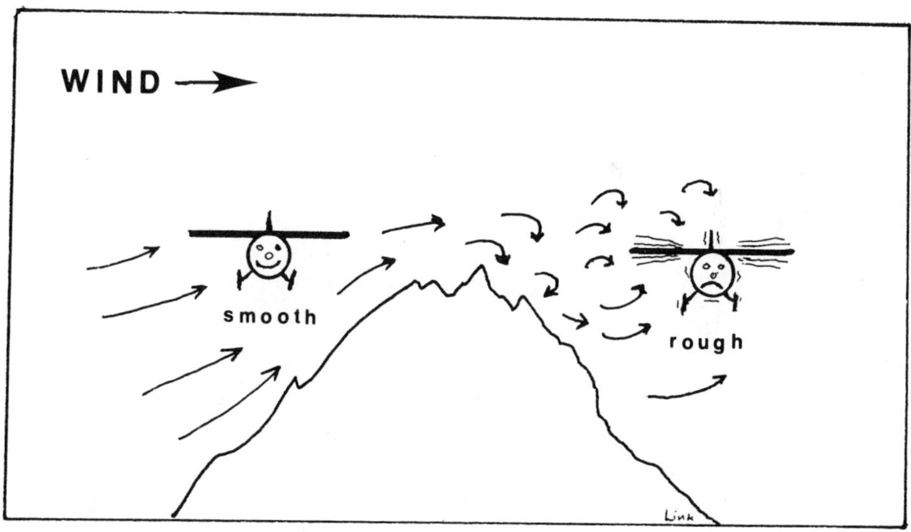

Turbulence near mountains usually *means wind is from the mountains.*

Wind in the Mountains

Wind is also important in new ways in the mountains. Various charts and diagrams are often shown that make wind seem more complicated than it is. Wind flows around and over obstacles just as any other fluid would, and if it is thought of in terms of water flow it may be easier to visualize it.

Wind will pass most easily up or down a valley between mountains and, just like a current of water flowing in a narrow place in a stream between rocks, it will flow faster. If wind flows over a mountain it will become turbulent just as water downstream from partly submerged rocks will become turbulent. In fact, one way that you can tell the direction of the wind in mountainous country is by the presence or absence of turbulence.

Mountain Turbulence

Turbulence around mountains can become particularly unpleasant; it may appear suddenly and unexpectedly and is often of a more violent and less regular character than in flat lands. Wind often turns into severe up and down drafts in mountainous country. The best way to avoid danger from this is to have lots of extra altitude. Rules are often given stating a specific amount of clearance to allow when crossing mountains, but actually the amount needed for safety varies with conditions. On a calm day the tops can be just cleared; while more and more wind calls for more and more clearance. Be alert for different wind conditions as you cross to the other side of the mountain. Sometimes it is best to cross the peaks at an angle so that if unexpected conditions are suddenly met, it takes only a partial turn to head the plane back towards lower terrain. If the wind is severe, it is a good time to keep out of the mountains completely. Winds can be especially unpredictable and treacherous in canyon areas and places where mountains come together from different directions. One of my favorite out-of-the-way flying places is Mexico's Copper Canyon, which is larger than our Grand Canyon. However, once wind comes up, I retreat to flatter land as soon as possible.[(47)]

Time of day can make big differences in flying conditions in the mountains. The usual recommendation is to fly in the morning and quit at noon until late in the day. This is a good general rule, but like most general rules it is not always applicable. The reason behind it is to fly in the cooler and smoother air early in the day to avoid density altitude problems and also to avoid turbulence which often builds up later in the day in the mountains. This turbulence can get really unpleasant and can cause early fatigue as well as roughing up the air dangerously and increasing bad wind conditions near landing spots. Frequently during very good weather, mountains often develop large cumulus buildups in the afternoons which progress into thunderstorms. On other days, the air stays calm and it is good flying all day.

Mount Shasta, California. *One of the Cascade Mountains which all look equally spectacular from the air. On a calm day like this we could easily fly close enough to wave at skiers, without any turbulence or other air problems.*

Keep It Light

Keeping the plane as light as possible is important in the mountains. Every pound that the engine doesn't have to lift gives that much more power for fighting downdrafts, climbing over ridges and other safety maneuvers.

AVOID MOUNTAINS IF POSSIBLE

When planning a route in mountainous country, try to avoid mountains as much as possible. Often a very slightly circuitous route following a valley instead of going over dense, rugged terrain, may take only a few minutes more flying time. If the route does have to cross mountains, plan it so that it is near valleys as much as possible—again a small deviation from the straight line can often make the route cover much safer terrain. Try to keep this in mind when looking over the sectionals in laying out the route. On a recent trip to Bella Coola (see page 236) my flying time was 17.3 hours. The direct straight-line route would have taken 16.5 hours. This shows how the addition of only a little extra time enabled me to avoid hostile mountains and very desolate country and stay over used roads and relatively settled areas. Incidentally, mountainwise, sectionals, WACs and ONCs show elevation contours and other potential hazards and are better than using IFR charts.

ARE THE AIRWAYS SAFER?

Are the airways safer? Some very experienced fliers always follow airways and feel that this is the safest way. They point out that the airways were established after long study and investigation, that they make search and rescue effors more efficient, and that there are more frequent airports along most airways. This is all true, but for going between two small off airways strips, you can often find routes that are much shorter than via the airways and sometimes cover safer ground than the airways route. Often too, just getting to an airway may require covering some hazardous country that could be avoided by forgetting about the airways. Airways are normally served by VOR's; but except for this, navigation may be easier off them. Using highways, lakes and peaks for checkpoints may be easier than following an airway over unremarkable terrain. Also, traffic tends to concentrate on the airways. Now that most commercial traffic is up in the flight levels away from VFR territory this is not so important, but in some congested areas it can still be a problem.

WHAT'S IN AN HOUR

Flying experience is usually measured in "hours" of flying. The FAA and insurance companies specify a minimum number of hours for various categories. Mountain flying rules typically specify 150 hours as a minimum requirement for mountain flying.

In reality, an "hour" is a very inexact unit of measurement. Lots of low time pilots have a natural proficiency that makes them better pilots then many high time ones. The experience value of an hour can vary greatly too.

Flying with the family using airline manners for many hours gives only a fraction of the training value that just a few hours of precision maneuvers give (but these would quickly have the family grabbing for airsick bags and resigning en masse from any future participation in flying). The point is, in mountain flying it's flying ability and understanding of potential mountain problems that count, rather than any specific number of "hours."

Keep Yourself Psychologically Uncommitted

Mountain and wilderness flying have taught me the frequent value of keeping an attitude of curiosity and using it to go out and *gradually* test situations. When we learn to fly, we seem to be taught an attitude that before taking off we are supposed to know quite exactly what route we are going to take and what the conditions are going to be like along this route. Up to a point this does promote flying safety, but if adhered to too rigidly it sometimes can actually decrease safety. It can make us develop a feeling that if our choice of route turns out to be more hazardous than we planned or if the weather becomes worse than we expected, we have somehow "goofed," but if we can muddle along through and make it, that will show that we were "right" after all. I flew for a long time before I found out that I could call en route and request a change in my flight plan route without being criticised. (After even more experience I finally learned that I didn't even have to *ask*; I could *tell* them I was changing it.)

In mountain and wilderness flying a *flexible* attitude can add much to safety. Charts often will not give you a completely accurate advance picture of the terrain you plan to cross and the usual unreliability of weather reports is accentuated by the scarcity of reporting stations in wild areas. So don't ever feel that you have to decide irrevocable beforehand what route you will use. Pick out the most promising route for where you hope to go but evaluate it constantly as you proceed and don't hesitate to change the route if the terrain or weather conditions become worse than you had anticipated. The point for such a decision will naturally be a variable thing for each pilot and each situation.

I, obviously, fly to many isolated places and over a lot of very rough terrain, but even so I am really a chicken and I very rarely keep flying when I can't see some spot on the ground where I think I could probably land without injury to occupants and with minimal wrinkling for the plane. I try to make an advance decision as to how long I will fly beyond the last such safe spot. This will depend upon how much altitude the weather will allow, how stable the weather appears, what the winds are like and how much further it is to a spot ahead that I *know* could provide a safe landing. Many of the interesting and isolated places that I have flown to were reached only because of an attitude of "I will look the route over and see." Often routes that looked bad on the map or were advised against by other pilots have turned out to have very good emergency landing spots all the way. At other times I have tried routes recommended by others only to say "No way!" and return. If you have this attitude about mountain and wilderness flying it will keep you safer. Don't be afraid to go and have a tentative look at a route, but keep the exploratory attitude – if it's not safe in your opinion, come back. Your retreat will not mean that you made a mistake in judgement – you just went to take a look. Nothing ventured, nothing gained! The time that you will make the mistake is when you feel that you have to *prove* that your judgement was sound by proceeding anyway.

This is about as spectacular and forbidding terrain as any I have flown over. Taken at 14,500 feet, the view is of the 13,000 foot Chugach Mountains (east of Anchorage) with icefields lying at about 8,000 feet. Yet from this altitude it would be an easy glide to the Glenn Highway which has many good landing spots on and beside it.

CHAPTER 8

THE BEST WILDERNESS AIRPLANE!

Oh! Wow! To recommend a specific plane or even a specific type of plane to a group of pilots is to step into a tiger's den! No subject seems to give more continual and heated debate among most hangar fliers than the "best" airplane for any purpose. Most pilots have long since decided upon their favorite plane and presumably are using one as close to it as they can manage. However, wilderness flying and camping do call for some unique specifications; and if a variety of planes are available, it is worthwhile to consider briefly the characteristics that are best adapted for this use.

Any airplane was designed by a series of compromises between possible alternatives. A fast wing for better cruise speed, or a slower wing for better low speed control. A large engine for speed and useful load, or a smaller one for economy. Two seats for economy and lighter weight, or four or more seats for larger parties. Two doors for safety and convenience, or one door for stronger construction and better aerodynamics. A high wing for shade and good ground visibility, or a low wing for better airport visibility and ground cushion effect. And so it goes.

Each pilot eventually, and, it seems, often irrevocably, decides which characteristics are most important to him and chooses the plane that most nearly fits his requirements. Others may have bypassed the choice because some specific plane is available to them. Often more emotion than reason seems to go into the choice.

For wilderness flying and for camping the following characteristics would seem to be most suitable:

A High Wing. For better ground visibility while searching for and evaluating wilderness strips. For shade and shelter while camping. For the usually associated two doors which are very convenient for camping. For the greater distance of the wing from the dirt, rocks and bushes that are often encountered off the pavement. For its usual spring type landing gear which is better on rough fields than the oleo type gear attached to a low wing.

To soothe the low wing advocates who are probably seething by now, a low wing also has advantages for camping. The low wing (properly protected) makes a good camping table, with a tarp thrown over it, a fair tent. Gas supply is easier to check. Stability in windstorms gives greater peace of mind in sometimes precarious tiedown conditions.

A Relatively Powerful Engine (or a relatively high power to weight ratio). It is obvious that a more powerful plane is safer for fighting downdrafts in the mountains, for getting to altitude quickly from valley strips next to high mountains, for being able to cross mountains instead of having to stay in passes, for climbing out of potential box canyons (where you shouldn't be anyway--but it does happen). The increased load capacity is valuable for the amount of gear that is often needed for wilderness trips. The only negative factors are the increased costs of acquisition, fuel and maintenance.

Low Octane Fuel Capability. A variety of octane ratings may not be available in out of-the-way places. One of the many reasons for the popularity of the Cessna 182 is its relatively powerful Continental 047 engine that (up until the 1977 models) burns 80 octane. So, at a Mexican beach strip, or anywhere else in the world, if the fuel is aviation fuel (and has been properly cared for) – you can use it.

A Fixed Landing Gear comes out slightly ahead and only because it is stronger than many retractable gears. Otherwise the greater speed of the retractable, plus the ability to put it down with a smooth bottom in some situations is better.

Taildragger Or Trigear? The taildragger comes out way ahead if only because of its greater prop clearance from rocks and dust and the absence of the tender nose wheel, (actually, the nose wheel is quite strong but what it is hitched to will often bend easily in rough field conditions). The taildragger's main disadvantage is the greater skill needed in ground handling. (If one were to sleep in the plane, the up angle toward the nose might be bothersome.) Sweden, long noted for pioneering spectacular conversions, now sells a light STOL plane that can be quickly converted between taildragger and tri-gear.

Single Or Twin? Another hot potato, and no need to enter the argument except to point out that the frequent shortness and roughness of wilderness strips favors the lighter single; while the usual advantages of two engines are increased in flying over wild country. The usually larger inside dimensions and useful load of a twin would be helpful in trips involving camping.

Wheel Farings. Pants, or skirts, on wheels look nice and cut down wind resistance a little. They also make it quite difficult to properly inspect brakes and tires. For the wilderness they are best left at home. They are ankle-bangers when you are working around the plane, and in mud and thick dirt they clog up easily and may create irregular and unplanned braking action on both landing and takeoff.

Floats. A floatplane is wonderful in wet wild country because it extends the possibilities of locating real wilderness to almost infinity. Serious are the disadvantages of the extra weight and loss of speed, especially for the long distances usually traveled in getting to wilderness regions.

Skis. For a skier, or those otherwise interested in getting into snowy country, skis are of limited use. Using them, and especially knowing how to select safe ski landing areas, takes very special skill and training.

Helicopters. A chopper should be the ultimate for looking for and landing at small, isolated spots in the wilds. But they rarely seem to be used for this, probably because of the high operating costs.

These are some of the major items, and if the selection of plane is flexible they should be considered, taking in mind past experience and capabilities as well as the type of flying that will be done.

That Gas Drip

Some high wing planes, particularly certain Cessnas, have a fuel system vent in one wing that will drip gasoline if the tanks are anywhere near to full, especially on hot days or if the plane is parked with even a slight tilt to this side. Even though a lot seems to run out, apparently the actual gas loss is never very significant. It is a problem, however, if one forgets and stands under the drip, or leaves anything under it. It is right over the tiedown point, so if anything important is used as an anchor it may get a gasoline soak. Nearby campfires or stoves could be a problem too. (I rarely have a complaint about the design of my Cessna 182, but this is one of them.) Living with the problem is easy if it can just be remembered. Park on a level spot or, preferably, one where the vent side is slightly elevated. If this can't be done, dig a small hole in front of the opposite tire and roll the plane into it. Taxiing up on a small sloped rock with the vent side wheel also helps. Remember that it will be necessary to taxi away from these aids when leaving, so keep them as small as possible.

CHAPTER 9

ASK LOCALLY, BUT BEWARE

Articles on travelling by lightplane often suggest that when in doubt about an unfamiliar area it is wise to ask some local pilot who knows the area. This can be excellent advice, *if* the one you ask is a competent pilot. To judge this is sometimes harder than the problem originally inquired about.

Think of some of the turkeys at your local airport. What kind of advice would a stranger get from some of them? Some of the most self assured and authoritative on the ground, who would most impress a stranger, are the least knowledgeable and competent in the air.

Several times I have been given advice by local pilots that was significantly erroneous. The errors varied two ways: some described a situation as being more dangerous than it was, while others represented conditions as being safer than they were. The latter type of error is, naturally, more dangerous.

I once asked a seemingly competent pilot working for a large reputable flying service adjacent to a large wilderness area about a 70 mile flight across it. "Nothing to it," he said. "We fly that way all the time. There are more strips under you there than most anywhere else." Well, he was right about the number of strips: They were small Forestry Service strips, but between the strips was nothing except rugged high mountains, some so high that gliding to any of the strips was, for a good part of the trip, impossible. Fortunately, and thanks only to our trusty bird, we got across without incident. Another time, while staying overnight at a small airport in British Columbia, I asked a local pilot about a route I had planned for the next day. He advised a much more circuitous route following several mountain highways and said that the stretch I had indicated had "No place to land anywhere." He was flying a small high powered taildragger and one would expect that he would recognize a minimal landing strip. Yet somehow, I felt unconvinced by his advice. My indecision was resolved at take off time next morning. His advised route was closed by mountain clouds and low fog. Mine showed adequate ceilings over its lower terrain. I took off to look it over and found that I could cover the whole route and always be within range of some spot where I should be able to put down even my heavier trigear and at least walk away. The next year, I happened to fly over part of his advised route and felt that, at least for me, it was considerably more hazardous than the way I had gone.

Once weather forced me to spend a night at a small airport on Vancouver Island. I met a local pilot who sounded like a very competent flyer with lots of experience in this region. Learning of my interest in finding wilderness strips, he suggested a very picturesque one on a nearby small island.

The next day, I went out and looked it over. It was really only a wide spot on a straight stretch of a little logging road. There was a parking spot, tiedown area and even a rudimentary hangar of sorts; so it was definitely an airstrip. But I wondered what kinds of planes had been using it, because its width, cut out of a thick stand of tall evergreen trees, was barely enough to clear my wingtips. I have enough things to think about in my landings, without having to watch my wingtips to dodge tree branches, so I passed this one up!

Another source of unexpectedly unreliable local information can occasionally come from the Flight Service Station. We all have had the experience of grossly erroneous weather reports where the weather specialist was obviously not looking out of the window, but instead relying on his teletype reports which sometimes seem to come via satellite from Istanbul. These reports can be wrong both ways, either describing destination weather as unflyable and when you finally arrive the next day you are told that the weather had been beautiful; or the report can indicate good weather which turns out to be no so, as you get near the area.

But there is another type of problem occasionally run into at the Flight Service Station; the local super-expert who goes way beyond the call of duty and beyond the amount of interpretation of data that you need. Often he is obviously not a pilot, but he is an expert on flying conditions and will go much further than he validly can in interpreting his reports. His implication is that, as a mere pilot, you do not have his understanding of meterology. Occasionally he will be overly optimistic, but excess pessimism is his usual bag.

When you first begin flying, you follow his advice; after all he is on the business side of the counter and he is wearing a pretty prestigious hat. The only harm done is that you have an unnecessary delay or that you stay home when you could have safely gone. But with more experience, you begin to recognize this psychological type and you learn to interpret the weather reports for yourself and to suspect when there is excess translation into them that is not warranted by the facts. You thank him politely for his help and proceed anyway. Now, the only problem is your passengers. They are still impressed by his position and are skeptical of you for thinking you can outjudge the expert. They will be nervous all the way. And, if you should have misjudged him and have to stop prematurely for weather – you have had it, as far as your status as a pilot goes. The only thing to do is to go to the briefing station alone. Or, if communicating by air, wear earphones and keep the speakers turned off. (Keeping speakers off also helps when you have the sort of passengers who turn pale when they overhear adverse reports for distant stations that do not apply to you.)

So – get all the advice and data you can about your flight and conditions along the way route, the more the better. But make your own evaluation of the situation from the actual facts that are available and learn to be wary of extravagant interpolations of the reported data. When getting opinions from local pilots, always pay attention to their ideas, but still remember that you, even though a stranger, may have had more general experience and be able to give a better evaluation than the local pilot.

But That's Life

All this is, of course, just part of what happens everyday in other parts of our lives. We are constantly being shown and taught things that are either right, wrong or in between. The successful person is often the one who can figure the best path *for him* through all this conflicting incoming data. And it's usually not absolute either; what is good for one person or one time, is bad for another. This is what makes hangar flying such an interesting part of flying life.

It was a little ironic that the first advertisement for this book in a flying periodical was separated by only one page from an article advising against flying a lightplane into one of the regions that I recommend (with reservations). The writer describes visiting this area on land and seeing only one airport which was below his standards. (Actually there are other very serviceable ones nearby this one.) He also describes seeing a Mexican charter plane with badly dinged prop and leading tail surfaces which indicated to him that the quality of the strips in this area was poor. (To me, this could indicate the condition at only one of the strips that had been used; it doesn't, by itself, rule out *all* of them.)

The point is that here is one of our most experienced and respected aviation writers who honestly feels that this area is not safe for plane or pilot. If I had read this article first I probably would not have tried to fly there. But fortunately, I had flown there first and found that this area (when weather conditions are right) is safe, at least for me. This is a very good example of why for wilderness flying you should not completely believe anything that you hear or read in flying articles, in this book or anywhere else. You have to decide *for yourself* what is safe for you and your plane. (Page 51 shows how you can evaluate such situations without getting into unsafe flying conditions.)

CHAPTER 10

MULTIPLE PLANES

TWO HEADS OR ONE • WHO IS BOSS?

Two or more planes flying together add a degree of safety to any long flight and especially so in wilderness flying. If one plane has to make an emergency landing, the other will know it and rescue operations can start immediately. Landings at deserted rough strips can be done first by the most experienced pilot or by the best plane for this type of operation. The following planes can then be guided by radio from what the first pilot finds on the ground.

The value of this procedure was illustrated to me once when I was flying up the west coast of Baja together with another plane. The usually excellent Baja weather let us down and it was necessary to land while we still could. We were over a little ranch strip that I was familiar with, so I landed first without any problem. My friend couldn't seem to get his plane to stop flying and had made two go arounds when I happened to look over at the dilapidated wind sock and discovered that the wind

had completely reversed itself since I had landed. I radioed this discovery to my friend who reversed his approach and then landed easily. The story had a happy ending because as we stood in this deserted spot under our wings in the rain, wondering what to do, a large covered truck drove up which turned out to be from a nearby orphanage. It seemed that three passes over the orphanage was a signal that someone was landing to visit it and my friend had made the three passes, without knowing it, during his landing attempts. We eagerly accepted their invitation to come over to the orphanage to get out of the rain and we ended up spending a very interesting night there. The girls helped to prepare the gigantic family style meal and to bed down some of the younger orphans, while the rest of us helped plug holes in the leaky roof, etc. It is run by a young American couple backed by a small religious organization in California. The orphans are darling (we wanted to take several home with us) and the work that they are doing with them is heartwarming to see. There is a great need for help either financial or labor or skills of almost any kind, and any pilot who wants to stop by to help will be warmly welcomed and have a rewarding time. (See page 234.)

We were given rooms in a building they called the "barracks" which was very adequate but a bit leaky (but then, rainy days are rare in that country). The architectural layout of the "barracks" puzzled us and finally the next morning we found out the real origin of the buildings. Just before the second World War some enterprising Mexican business men had planned a dam on the adjacent stream to make a fishing lake and started the construction of a resort beside it. The war ended the effort but enough building had been done that it later was adequate for the orphanage to take over and use. The "barracks" was originally designed to be the workshop of the resort's call girls, which explained it's somewhat unique design. My wife still enjoys telling about the rainy night she spent in a Mexican whorehouse!

TWO HEADS OR ONE

For trip planning and for evaluating flying situations, two or more heads may be better than one, and the combined experience of several pilots may lead to a better planned trip. Of course, there is the old story of a committee building a horse and ending up with a camel – combined plans may lead to a compromise procedure that is not the best for anyone. Flight planning is largely a personal thing and each pilot should make plans that best fit his own proficiencies and experience. It really just depends on personal preferences. Many pilots like to have others along for company and for consultation in questionable situations. Others are more the lone wolf type and are happiest when they can run the whole show. Then too, with several planes there are that many more mechanical components that could give trouble and slow up the trip.

Before a trip with more than one plane, there should be a discussion of traffic rules. (Amateur formation flying has often ended in disaster.) High wing planes should always try to be on top and low wing planes underneath.

WHO IS BOSS?

A somewhat analagous situation arises when there is more than one pilot in a single plane. Who is going to play what role in running the show will usually be worked out more or less naturally by interaction of the personalities involved. But in any flying situation of this sort, whether in the wilderness or elsewhere, there should be a definite understanding, before the flight starts, as to who will be in command and fly the airplane if any emergency arises. Accident reports are full of incidents where, when a situation became suddenly and unexpectedly critical, an accident occurred only because for several vital moments the plane was not being controlled while the two pilots inside were working out who

was going to take charge. Presumably it should be the pilot in the left seat, but he may not be the most experienced pilot. A plane owner may feel that he knows the plane best and should fly it; while another pilot may be much more experienced in general and feel that he is best qualified to take over.

Once I was flying my own plane on a camping trip from Fairbanks to the North Slope. Beside me was an experienced local pilot who was very familiar with the area and who had been directing me as to the proper headings, etc. Just beyond Anaktuvuk Pass we flew towards a steadily lowering ceiling of arctic fog. The local pilot, who had been commenting on flying conditions along the way, said nothing, so when we were down to about a 200 foot clearance I finally said, "I think we should turn around" to which he quickly replied, "I've been wondering how long you were going to keep on in this stuff." We had each been assuming that the other was in charge of such decision making and waiting for the other to speak. Fortunately, no real danger was involved, but it was a good example of why, *before* the flight, there should be an *expressed* decision of who will be pilot in command if anything should happen.

Gradually lowering ceiling described above in Brooks Range. *It is very flyable here but beyond Anaktuvuk Pass it met a mass of artic fog and became IFR.*

CHAPTER 11

TIEDOWNS (THE JUDGE MAY NOT UNDERSTAND FLYING)

DO IT NOW! • DON'T DOUBLE PARK!

Many off pavement strips have rudimentary or, probably, no tiedown facilities. Many times, other pilots will have left in the tiedown areas items that can be used – large rocks that are irregular enough to tie a rope around, heavy pieces of junk metal, etc. (These can also be hazards to the prop or gear when taxiing for parking. If there is tall grass or the area is otherwise hard to see well, stop the engine and get out and look around carefully on foot.)

Low wing planes need less tie down than high wing planes, and all planes need a lot less tie down than is usually supposed. But in the rare case that a tie down is needed, it is really needed and an upside down plane in the wilderness is indeed a predicament.

An emergency tie down kit should be carried for times when no heavy objects are available. Various devices are popular for this – metal stakes, corkscrew type rods, etc. Some pilots affix long pieces of lightweight aluminum angle iron across the horizontal surfaces while others tie on extra clothing or blankets to spoil the lift of the wind.

One of the most functional and easiest to carry, which will work in any type of ground, is three sturdy burlap sacks and ropes to tie them to the plane with. They are filled with rocks, or other heavy things, under the principle that enough weight to hold the plane down can't be carried at once, but the accumulated weight of a number of lighter pieces that can be carried, will be adequate. These will probably not be needed very often, so stow them in one of the less accessible corners of the cargo area. If the three sacs are put in a medium sized nylon "stuff sack" of a distinctive color they will stay together and be easier to find. The ropes should be stored separately, but together, because they will be used more often. Your own ropes should always be used. Don't depend on a rope left at an unattended strip. They often look thick and strong, but many a plane has been lost by the unexpected parting of a strong looking but actually weather deteriorated rope.

Putting chocks under both sides of all wheels helps a lot too. There are small lightweight aluminum chocks available that are very lightweight and efficient; but these are just one more thing to have to carry and keep track of. Almost every strip has loose rocks around that will work just as well.

Clinton Creek (Cousin's) Strip *outside Whitehorse, Yukon Territory. At isolated strips most anything may be around to tie down to. (This is a good camping strip, see page 237.)*

DO IT NOW!

Get into the habit of tying down and chocking the plane immediately after you have decided to stay for awhile. If you don't secure the plane immediately and start looking around, it is easy to get distracted by interesting things and neglect the tiedown. In strange places the weather patterns may be quite unfamiliar, and in some places very high winds can come up very suddenly.

DON'T DOUBLE PARK!

When tying down, or even parking for a short time on a rural strip, always be sure to move the plane as far as possible from the active part of the runway. Even when the area seems to be permanently deserted, it is surprising how often the only other plane to use if for a month will arrive while you are there. The next pilot to land may need a lot of room.

Once I was following a Baja-1000 race by air. I had landed at a small strip beside a pit stop and had parked in the crowded parking area, next to, but well clear of the active runway. Very soon another pilot lost it on landing. He hit the plane next to mine and shrapnel from parts of his plane hit my plane and me. The landing pilot claimed that his problem was caused by the plane next to mine starting to pull out on the strip as he landed (actually not so). A long controversy between them ensued, finally ending up in court.

The jury trial was a sobering lesson for me. The judge, jury and most of the attorneys knew nothing about planes or flying. The landing pilot made up fantastic stories about flying and landing which would have had him very quickly hooted out of any hangar flying session. He delivered these wild tales in a very confident and authoritative manner which was

obviously impressing the jury. Unfortunately, the ground rules of the courtroom made it impossible for anyone who knew better to challenge him. I was suspect for a time, on the theory that my plane was an obstacle to the runway, and I was barely saved and only by a photograph that happened to show the berm at the runway edge, well ahead of my plane. The case did not come out the way it should have.

This experience showed me the advisability of doing everything possible in every flying situation that would indicate to the non-flyer that you were extra careful. Extra radio calls, parking way off the runway, stopping for other taxiing planes and every other often unnecessary thing that you can think of to show a non-flyer that you were a super careful guy can make a big difference in what might be a very important decision for you.

CHAPTER 12

THE EMERGENCY LANDING

PRACTICE GLIDING • LAND UPHILL • HIT TWO TREES? WHEN IS AN EMERGENCY • DITCHING

It seems somehow appropriate that a book on wilderness flying should have a section on emergency landings. In many people's minds they somehow seem to go together. The rare emergency landing in settled country may make the newspapers for one day; but an even rarer emergency landing in the wilderness often becomes a prolonged epic that people remember.

All phases of the emergency landing should have been well covered during basic flying training; and, again, this is another phase of flying that needs actual practice and can only be partially learned by merely reading about it. Before a wilderness trip, one should review the procedures and techniques that he was taught for use in an unexpected landing. If you have been a flyer solely in the flatlands, there may be some new concepts that should be thought about. If you fly in the more mountainous areas of the country, you may be already familiar with most situations that you can run into in the wilderness.

In case of engine problems response to engine restart procedure should be nearly automatic (mixture rich, fuel pump on, change tanks, or whatever for your plane). As you are doing this, slowly put the nose up to reduce speed without altitude loss to the best gliding speed for your plane and at the same time review the surface wind direction (which you should always know almost subconsciously) and locate the most promising reachable spot for an off strip landing (which should also be a continual subconscious habit when flying in rough country). If there are two or more equally promising and equally distant spots, go for the downwind one as the plane will cover more ground gliding that way. In your spare moments and while you still have as much altitude as possible try to get off a May Day message on 121.5 or whatever frequency you have been using. Hopefully you can give a pretty good estimation of your location. Maintain the best gliding speed, get all seat belts and, hopefully, shoulder harnesses tightened well. Many wilderness pilots try to pack their planes so that foam mattress rolls and other soft items are easy to get to, to put in front of passengers' heads and chests (and yours too at the last moment if the spot looks bad). Think about fences across fields, wires that you can't see that will be running between poles that you can see. Also be ready to lean the engine to stop, if it is still running at all, and turn off the ignition and master switch at an appropriate moment, while opening doors with your other hand. Put landing gear in the position you feel is most appropriate for the terrain that is coming up. Some super cool types also try to get the prop, if it's two bladed, in a transverse position with the starter to minimize prop and engine damage. Otherwise, try to relax as much as possible (sic) so that you can give your all to the dead stick approach and landing that, hopefully, you have kept current on.

Well, that's the Emergency Landing section which seems appropriate for a book on wilderness flying. But it shouldn't really be here at all, because anyone who flies anywhere, and especially anyone who flies in wilderness areas, should have all this memorized and rehearsed and frequently practiced. Not as someone else may write about it in a book, but as he can best do it in relation to his own experience and proficiency and the needs for his particular airplane.

PRACTICE GLIDING

Every so often when you are flying, preferably when flying alone or with another pilot, practice evaluating how far you can glide without power from your altitude. Pick out a spot that you think you could glide to and pull the power off, set up your maximum gliding distance speed and see how far you get. Power should be put on, to keep the valves warm, for a few seconds every minute or so but this will make only a small difference in the distance covered. If you do this in the mountains, you will probably find that for a considerable time you will be doing beautifully; you are covering a lot of the distance and with seemingly very little loss in altitude. Suddenly, however, it appears as if you are too low. Higher spots of the terrain are quickly getting too close and you realize that you won't make the chosen spot. Understanding of this phenomenon of suddenly finding that you are too low at the last minute is important so that you will be able to allow for it and have secondary landing possibilities picked out rather than having to put it down in a rough place at the last moment.

LAND UPHILL

In a forced landing in rough hilly country, it is occasionally best to land up the side of a hill. A spot on the side of a hill or mountain may be smoother and have less obstacles than the flatter areas nearby, and in an uphill landing gravity can help slow you down. This requires a lot more flare than you are used to making and it is something that is difficult to effectively practice.

HIT TWO TREES?

It is sometimes recommended, when there is not much room for the emergency landing in wooded country, to hit two trees simultaneously – one with each wing. Theoretically, the wings come off and in doing so absorb a lot of the landing momentum. Well – maybe, if it's that or nothing. But I have seen wings forced fatally through the sides of the fuselage. Also, if both of the trees are not hit at the same instant, a massive change in direction will occur. So try it, but only if it seems to be the only way to slow the plane.

If a forced landing ends up with your plane upside down, you will probably be hanging there by your seat belt. It is an unnatural and unpleasant position, to say the least, and the natural instinct is to correct it just as quickly as possible. However, before releasing the seat belt, be sure and get a firm hold on something solid or brace yourself against the roof (which is now the "floor"). Many fliers have survived an inverted landing completely unharmed, only to sustain often severe head and neck injuries when they release the seat belt.

WHEN IS AN EMERGENCY?

Most flying instruction and literature cover the emergency landing as if it would be a definite situation about which there would be no question. Presumably the engine will suddenly make atypical amounts of noise or quietness and there you will be, way up there with no power, having to glide to an emergency landing. Well, often it is like this, but probably much more often, it is not. Actually, one of the most difficult and most critical parts of emergency landing procedure may be to determine *if* you need to make one. Rather than an obvious inability to function adequately, the engine may show varied degrees of weakness, and it is often a very hard decision whether it's better to put it down at once in a questionable spot or to try to nurse it along to a better one. It is a decision that you will have to make at the time depending on the conditions then, but it's good to think about such possibilities ahead of time. The panic feeling will be a little less if it's not a sudden completely unexpected situation.

When I started flying, I had always supposed that a decision for an emergency landing would be a clear cut one. The first time that I had carb ice I did not recognize it and with the plane running at about 40% power I started checking out emergency landing spots. I could not find one I was satisfied with at first, so I kept flying and looking and suddenly realized that the plane was still flying and not getting any worse, and I made it to the next airport. Yet, if my first emergency spots had looked adequate I would have landed there and no one would have considered there was any other alternative.

A point that should be repeated again is that flat country fliers should study, review and think about the different emergency landing

My son was flying this Arctic Flying Club 182 in the Brooks Range when it gradually lost power and could not maintain altitude. By maintaining minimal flying speed *and flaring up a very rocky hillside, he managed a landing in a very hostile spot with one tiny cut, the only injury to the four occupants.*

spots that may be found in the wilderness. It also never hurts to be reminded once again of the fact that, almost no matter what, if you can land your plane right side up at the very minimum landing speed you are far, far better off than stalling it in. All survival instincts tell you to pull the nose up to get slower and to keep away from that ground as long as possible, but instinct in wrong; and the vital necessity of ignoring instinct in this incidence just can't be repeated often enough.

Actually, unless there is considerable altitude available, many emergency situations proceed so suddenly and unexpectedly that you don't have time to do anything except sit there and hope – for a very brief moment. This is another reason why emergency procedures should be rehearsed until they are practically reflexes.

DITCHING

A discussion of emergency landings is not complete without mention of the water landings. Much has been written about the actual techniques for ditching and it does not have to be repeated here, except to suggest that one of the most complete, clear and concise descriptions is that in the Airmans' Information Manual, Part 1.

If there is a choice of putting it down on land or in the water it will depend upon how good a spot is available on the land. Water seems softer and more gentle than land, and even with all the technology available to it, NASA found it best for the astronauts' reentries. But when hit at flying speed water is actually quite hard, resistant and damaging. If there is a reasonably good spot on land it would probably be preferable.

If a plane can be set down on land without serious injury, and without burning, there is time to force jammed doors open, pause to get over a temporary stunning, to gently remove an injured occupant, and to get supplies and gear out of the plane. In the water, the time for these things is limited because most all planes are noted for their rapid tendency to sink.

In much of the north country, the water (and often the air too) is so cold that even if a pilot gets quickly out of a ditched plane and to shore unscathed, he and whatever gear he may have been able to salvage, are so soaked with the near freezing water that survival may still be tenuous.

So I always try to get enough altitude to glide to land, or else fly near the shore and around very large lakes or ocean bays so that the land-or-water option will be available.

Flying in the Caribbean made a great change in my attitude about flying over water. Up until then my experience had been mostly over the relatively cold and very isolated shark-infested water of the Sea of Cortez or else in the very frigid and almost equally isolated waters of Canada and Alaska. Survival in these waters is tenuous because of their low temperatures and also because *if you had to ditch,* especially in Mexico, no one would probably know you were there, and even if they did, rescue facilities are few and far away. When I first read the AOPA booklet on the Bahamas, I was quite skeptical about their section describing the relative safety of overwater flying there. But when I flew there I found it seemed to be true. The water is relatively warm which makes it somehow more benign and actually does increase survival chances. It is also very shallow in many spots and even though it is usually too deep to stand up in, it makes it *seem* safer. There are also innumerable uncharted rocks and small islets but the main safety inspiring feature is the good radio coverage and nearby rescue units on shore. The U. S. and Bahamian radios cover the entire area, and often overlap and the impression is given that if a mayday message was sent out, help would soon arrive.

CHAPTER 13

SURVIVAL

ALWAYS TAKE A GOOD SURVIVAL MANUAL
SURVIVAL GEAR • DON'T RUSH OFF • SURVIVAL TRAINING

Survival is an ominous word, but in some ways a happy one. The forced landing has been survived and now survival is important. The speed and comfort with which the modern lightplane can cover wilderness areas makes it almost impossible to comprehend the extreme difficulty and the snail like speed of any other way of traversing the rugged terrain (which so often looks deceptively benign as seen from the comfortable plane). It is easy, therefore, to minimize the possible consequences and omit precautions that could be life saving. Even on routine flights that cross desert areas where a forced landing would mean 24 hours or more in desert, extra drinking water is often omitted. In potential cold areas a warm jacket is not taken. On a usual stateside flight, if one is forced down and not injured, these little omissions may cause extreme discomfort but usually no more than this. In wilderness flying, however, omission of necessary survival gear can be much more serious. Rescue may take a long time, ground weather conditions may be much more severe, and an overland route to safety may be extremely long and hard. Here, proper survival gear may well become exactly what it is called – gear necessary for survival.

ALWAYS TAKE A GOOD SURVIVAL MANUAL

A survival manual is very essential. It must be accurate and complete. Most wilderness flights cover a variety of terrain and the manual should cover them all – artic, tropic, desert and forest. One of the best is put out for the Armed Forces.[6]

Get one! But don't spend too much time reading it, unless this sort of reading turns you on. Just have it with you. If you ever do need it, you will then have plenty of time on your hands to read it and at that time it will also have more real meaning for you.

SURVIVAL GEAR

Also in the appendix are lists of required and suggested items of survival gear as advocated by the Canadian Government for its sparcely settled areas, and by the authorities for Alaska.[7] You will probably not agree entirely with their selection, but consider the items carefully anyway because some are legal requirements and all were selected by experts, based on a lot of actual experience.[8]

So select your survival gear with care. Get everything that you will need for every type of country you will be traversing. Get items that are as small and lightweight as possible and still do the work. Familiarize yourself with them and read the manual as much as you have the motivation to. Then pack them carefully but compactly and stow them in a back corner of the plane. This allows you to use the more readily accessible areas for things that you need more often. If you are on a camping trip you will probably automatically have along fairly adequate supplies of food and warm weather gear. You may want to add to the recommended items a few extra things for your own personal comfort and convenience, such as a pack of playing cards or some paperback reading material. It can be a long wait out there! (One reminder – never take a survival firearm – or any other kind – into Mexico. The chances of being jailed for this are greater than the chances of needing it in a survival situation. And the inside of a Mexican jail has poorer survival conditions than most anywhere else!)

DON'T RUSH OFF

If you have survived a forced landing, get away from the plane as quickly as you can until you are sure that it is not going to catch fire. (If it should, it can almost instantaneously become a great ball of fire.) After this, do nothing for awhile. Sit or lie down and relax. There is nothing that has to be done or any decisions that have to be made immediately. The tension during the forced landing combined with the probable trauma of the actual impact will make definite changes in body chemistry which take a considerable time to overcome. The overwhelming relief of the survival often numbs the whole body; you may feel just fine but in a little while this feeling begins to wear off and you may find you are not doing as well as it seemed at first.

This phenomenon was dramatically demonstrated to me when I made a hard landing at Gallup, New Mexico. (It was hard enough to do $7,000.00 worth of damage to the plane and put two of us in the hospital overnight! Ironically, it occurred on a wide, paved runway 6,500 feet long just after I had returned from an Alaskan trip with many a landing on very marginal strips. Officially the cause was a defective nose wheel, but I suspect that a complacency from over-confidence played a large part.) We quickly got out of the crumpled plane, and I was relieved to

find that I felt just fine. I was even running around taking pictures from all angles when I noticed that the contour of my arm seemed unnatural. X-rays later confirmed a fracture which needed setting and a cast for six weeks. Yet for a considerable time after the accident I felt perfectly all right.

So, if you have such an unfortunate experience, relax, rest, do nothing and make no decisions for some time, until you are in better shape to evaluate your situation. If you are in an isolated area, staying with the plane for at least several days is almost always the best. The plane is much more likely to be found than you alone are. Often parts of the plane can aid in shelter if foul weather comes up. The effort of carrying supplies away from the plane can use up energy that may be needed for survival later on. If you try to make it to a better location, be sure that you can find the way and always remember that things on the ground are much farther and harder to walk to than it seemed from the air.

SURVIVAL TRAINING

Many courses are available in survival training. Some adult education classes or university extension courses cover general survival situations. The Baja Bush Pilots have had a number of realistic weekend fly-ins where, under supervision, practice is done in living for 24 hours with only the survival gear brought in the plane. The AOPA has an excellent course with some thirteen hours of classroom and in-the-field training spread out over three days. Such survival experiences give excellent training and practice, and can also make a very interesting weekend, if you want to spend this much time at it. However, if your time and finances are limited, you should consider seriously if one of their actual flight training courses would not be of more practical benefit. There is hardly a pilot who cannot benefit from more flight instruction; it is something that can be used actively in nearly every flight, whereas survival training is effort spent for a situation that is really very unlikely to actually occur.

Survival courses must take a lot of time to cover different situations: arctic, desert, tropics, forests, and the ocean. In any actual situation, you will only need to know about one of these. Much of the information and skills that you may learn will decay quite rapidly if not used or practiced regularly. There would probably not be much retention for long of identification of many edible versus non-edible plants, for example, or of the details of rigging a snare, etc. If you are at all the outdoor type, much that is taught is already contained in your own common sense. One picture of a survival course shows the class out in a field learning how to shoot flares and to use signal mirrors. Do you really need to take a course in order to know how to use these effectively?

If you have not had much outdoor experience and feel uncertain about how you could take care of a survival situation, a course can give you a lot of valuable help, especially a boost in self confidence which is a large part of it all, Otherwise, an almost equally efficient approach is to just have aboard those survival items advised by the experts along with a good survival manual.

This is not meant to put down survival courses. It is only a suggestion on the ordering of priorities. If you have time and feel the need, they can give you vital information. But there are many with experience and aptitude for the wild country who can do nearly as well with a good survival manual if and when it is needed. For these, the time and expense of a survival course might be better used for further refining flying skills.

CHAPTER 14

PROPPING

RELAX • AN UNFORTUNATE COINCIDENCE

An understanding of how to safely start an airplane by turning its propeller is necessary for wilderness flying. In civilization there is usually help for starting a plane with a dead battery, but in the wilderness there is often none, and there may be no choice but to prop it yourself.

I have had dead airplane batteries from all of these causes: sudden battery breakdown; leaving the master on; using plane lights too long at night; a broken alternator ground wire connection; a balky engine; and there are lots of other possible reasons that I will probably have eventually.

Fortunately, because magnetos create the electricity to fire the spark plugs, an airplane, unlike an automobile engine, does not need battery current to run. But battery current and a lot of it is needed to turn over the starter and a dead battery does present a problem in the wilderness. If the battery is dead and the alternator or generator does not work, the radios and electrically driven instruments will not work. In the wilderness these are of secondary importance, but the starter is vital.

Some light planes have no battery and depend on propping to start. These have small engines which are easier to turn over with the propeller. But the heavier planes that are usually involved in wilderness expeditions have larger displacement, high compression engines that are much harder to turn over by hand. (On some planes such as the Cessna 175 or mid-1970 172's a gearing arrangement of the prop or starter makes it difficult or impossible to prop it.)

Propping is very serious business. If done successfully one is suddenly inches away from sudden, messy death. Propped planes are notorious for escaping from apparently adquate restraints and running off by themselves to chew up miscellaneous things they find in their way.

RELAX

A sudden, unexpected need to prop in the wilderness can be emotionally upsetting. You may feel anger at the mechanical failure, rushed by the unexpected delay, and be apprehensive at the possibility of being stranded there. This natural reaction helps cause many propping accidents, so pause, take a few deep breaths and slow yourself down. Besides, the success of the propping will be more likely if you can be calm and relaxed.

Do not prop if there is any other way to start the plane. If there are the facilities and the time, change or charge the battery. If a "jump" is available from another plane (awkward) or an automobile or truck (easier) do it. It is still a cumbersome chore and there will still be someone outside the plane potentially near the propeller, but it is much safer than propping. Use automobile jumper cables and, just as with mating two cars electrically, be sure that the polarity is correct (i.e. the plus terminals of both batteries are connected to the same cable). Otherwise expensive damage to alternator diodes and other components can occur. The voltages of the two systems do not have to be the same. If all electrical components (lights, radios, etc.) are turned off, up to twice the plane's voltage can be safely used for a brief start. In this situation disconnect at least one end of one cable as soon as safely possible after the start. Run the engine of the car at moderate speed so that its generator or alternator can help out its battery. But don't try to run a donor airplane engine for such a boost, one turning prop in the area is already one too many. If the donor voltage is lower than the plane's (e.g., a twelve volt plane and a six volt car), the results may not be too good, but it is worth trying, especially if the donor car can rev up its engine. If the starter stops moving – switch off! If the starter armature is not rotating, even the low voltage running too long in one segment of the wiring can quickly burn it out.

PROPPING

Master Off. Keeping the master switch off during propping protects the rest of the electrical system from voltage surges during starting. If there is an alternator, it will be very hard to turn it if it is in the circuit, and master off can keep this pull from being added to your already heavy burden.

Brakes. If possible have someone sit in the plane and stand on the foot brakes hard. If it is another pilot whom you know or someone else you can completely trust, you can call commands about switching on and off. If there is any question about such communications, do the switching yourself and keep the key in your pocket when you want switches off. At least be sure that the person inside knows how to stop the engine quickly, if necessary (as well as understands how to put on the brakes).

Secure It. You cannot be too sure that the plane is securely fixed in place before propping. Chock the main wheels, (but not the nose wheel – too near the prop). But do not depend only on brakes and chocks. Have a sturdy rope tied well to a solid object behind the plane. Tie it snugly – it takes much less to keep a plane from starting to roll than it does to stop it if it moves even a small distance before the rope tightens ($M=WV^2$).

AN UNFORTUNATE COINCIDENCE

Three days after I had written these precautions about propping, a friend of mine had an unfortunate accident which served to emphasize some of these points. As he was propping his cherry Cessna 140, it got away from him, nearly ran him down and chewed up the plane tied next to his for a total damage of some $7,000.

Wing of the plane described above. Note location of strut, a few more inches inboard and the fuel tank would have been chewed up with possible tragic results.

He is one of the most competent pilots I have ever flown with, young and alert with instrument, commercial and multi-engine ratings and much experience with many types of planes. He is named on my airplane insurance policy, and I feel fortunate to have him to ferry it for me sometimes.

Yet, in a minute of haste, he overlooked the following points when he suddenly found his battery was dead: External power for a jump was right beside the plane; experienced pilots were nearby and ready to help; the plane was secured only by two rather ancient chains, one to each wing tiedown ring and in vertical positions that gave little horizontal security; a friend inside the plane was not instructed how to put on the brakes or how to stop the engine.

Oh-Ho!

An interesting sidelight to this incident was that when he tried to report it to the FAA, they were not interested because he was preparing to taxi it only over to the wash rack. Apparently on the ground accidents are not reportable accidents *unless the plane is taxiing for takeoff.*

In the appendix there is a "check list" for propping with an extra copy that can be removed and kept in the plane. I keep a similar one in a plastic cover in my flight bag. When the time suddenly comes that propping is necessary, it can be a very helpful review and decrease potential hazards.

PART III
CAMPING BY LIGHTPLANE

CHAPTER 15

TO CAMP OR NOT TO CAMP

Many private pilots, even many with long and varied experience, do not realize how much they can increase the utility and enjoyment of their planes by the use of off-pavement strips and further by the use of camping. As most pilots know, the convenience and time-saving of a private plane often comes to an abrupt halt when a flight is temporarily interrupted by an overnight stay at an airport. Even on busy paved strips, it may take a little effort to locate a phone. Then there is a delay while the taxi or other ground transportation shows up. The drive to a motel may take some time. If one is particular about a motel, it is sometimes a little awkward as well as time-consuming to look at different ones by taxi. Once in the hotel, unless the organizing has been better than the average, something really needed will often be discovered to be still in the plane. Perhaps the maps left in the side pocket are necessary to review plans for tomorrow's flight, or a particular item of needed clothing, medicine or literature is still in the plane. This problem can be solved by taking everything out of the plane and into the hotel; but with the amount of gear often taken on an extended trip, this can be a time-consuming, laborious chore which may completely upset the carefully arranged packing that preceeded the trip. Dragging around all this gear, perhaps with frowns from the taxi driver, the need

for extra bell boys as well as the crowding of the hotel room, may put a little damper on the fun of the trip. There is always the chance of having something valuable stolen, whether it is left alone in the hotel room during meals, etc., or left in the deserted plane.

Then, in the morning, the whole process is reversed and more time and effort are consumed. If an early morning departure is desired, the distance separating the plane and the sleeping accommodations may give still more problems. Motel phones are sometimes turned off for the night and a taxi cannot be called, or one may not be available for the early bird. One can usually make advance arrangements for early morning transportation, but if, at the early prearranged hour, there turns out to be an unexpected temporary ground fog, or some other adverse weather development, the taxi may still arrive. The taxi driver doesn't watch the sky, only his meter. Even if the taxi is then postponed, the encounter at the door may have ended the possibility of any more sleep.

Conversely, if an earlier peep at the weather shows extra good conditions that should be taken advantage of, the ground transportation may not be flexible enough to get one to the airport at this earlier hour.

Contrast all this with staying with your plane overnight. You have saved the time involved in selecting a proper hotel room and getting to it. Any needed gear is in the same place that you are, without the effort of having had to move it. You have been able to watch it all night. You are right on the scene to observe the weather and you are able to take off as early or as late as you desire without being dependent on any ground transportation. If you have selected a proper spot, your night's sleep was probably better because of quieter conditions than at most hotels or motels.

When there is full camping capability aboard, it adds a dimension of safety to weather situations. Often, the take off area has safely flyable weather, but there are indications of questionable conditions along the route. If you are not dependent on fixed ground accommodations, you can start out and see how the weather is as you proceed. If it starts to decrease toward your personal minimums, you can usually find a spot nearby where you can set down and be comfortable.

Without camping gear aboard, your go no-go standards have to be more rigid. If there is much chance of unsatisfactory weather along the route, you will probably stay where you are until you can be more sure that the entire route will be satisfactory. If you do start out and the weather turns sour, you have more pressure to continue on, because your only alternative is to return to the starting point. Even if you 180, the weather may have soured behind you, too, so that the only way that you can get to a satisfactory stopping spot is to fly through bad weather.

Quite often, the questionable weather reported along the route does not materialize, and if you can safely start out and look you will end up getting through with safe conditions. Even if you have to stop part way and camp, you can often continue later from this point and end up by getting through sooner than if you had to do it all at the same time. Conversely, good weather predictions may be wrong, and camping capability will enable you to stop much sooner than if you need to try and find a spot with adequate RON accommodations. So, when the weather does not turn out to be as advertised – whether better or worse – with camping gear aboard you are safer either way.

All these advantages of staying with the plane are, of course, obvious to the travelling private pilot. Yet for many, the idea of sleeping by the plane seems to give mental pictures of discomfort, dirt and deprivation.

Yet, in actuality, an unbelievable new world of camping comfort and ease is available today. In just the past few years, outdoor equipment manufacturers have used space-age materials to create equipment that is ultra lightweight, quickly assembled and has a luxurious comfort undreamed of only a short while ago. With a really minimum of time and energy and with the addition of only very little extra weight and bulk to your gear,

you can set up a lightweight modern camp, a very comfortable and quiet room for yourself.

The benefits of camping were again demonstrated to me recently while returning home from visiting friends in northern British Columbia. Usually, we somehow think of camping as being done in idyllic surroundings while participating in relaxing recreation. This example shows the time and energy saving possibilities of camping even under much less than ideal conditions.

It was late in the day, darkness was coming on and the weather was beginning to go sour. On the way up, I had spotted a little gravel strip at an unused logging camp a few miles north of Prince George and was planning to camp there that night. Somehow in the fading light and weather I missed it and was in the control zone of the Prince George airport before I knew it. There was only about 20 minutes left before the Canadian VFR curfew time so it was too late to retrace my route to try and find that camping strip – so it had to be Prince George that night.

Camping is frowned upon at the Prince George airport. You can sleep in or beside your plane but tents or fires are not permitted. No one would normally want to camp there anyway because of a busy scheduled airline with occasional jet traffic all night. But that night I didn't care. I was tired!!! My gear was all arranged (or disarranged) for camping; to collect the necessary items to go into a motel seemed like just too much work! It would have taken a good twenty five minutes for the taxi to come out from town and an equal period to get back to the motel. I had my dog with me and I was not sure that she would be welcome at those motels, and, although she is trained to stay alone in the plane all night, somehow being away from home, tired and hungry and with bad weather approaching I wanted her for company that night. It was also getting rather late for any restaurants to be open by the time I would get settled in the motel in town and even if one was open it might be a long way from the motel. The plane contained comfortable, cozy sleeping gear, very adequate food that could be quickly prepared and it was all ready *now*. I tied down my plane in the grassy transient parking area, fortunately a considerable distance from the busy terminal. Rain was due and since a tent was not allowed I put all my gear under the wing and covered it with a tarp and made up my bed inside the plane (the rear seats had been left at home). I walked over to the weather office to look at the weather data, I washed up in the waiting room restroom, ate dinner in my plane, fed my dog, put in my earplugs against the noise of the jets that were scheduled to taxi by several times during the night and, after reading a little, was soon comfortably asleep. I awoke early in the morning to the patter of rain on the wings. A peep out the window showed that there was obviously no flying to be done for some time, so I went quickly and easily back to sleep.

I awoke several hours later to find the rain had stopped and that there appeared to be good VFR for at least 25 miles in the direction I wanted to go, although the weather in all other directions still looked bad. I walked over to the weather office and found that the reports for my route were still minimal but improvements were forecast for later in the day. Here was where camping *at the airport* paid off! It was quick and simple to load my gear in the plane and fly down to see what the weather was like further on. I had nothing to lose – if I couldn't get further than the good weather I could see, there were spots there where I could land, put up my tent to sleep or read. Local weather was very gradually improving, there seemed no chance that it would suddenly go IFR, so I could even have safely returned to the airport with nothing lost for the attempt except a little time and gas.

By contrast, if I had been in a motel in town, the pessimistic reports that I would have received by phone would have made it seem as if the only thing to do was stay at the motel. It would not be worth the gamble to pack things up, check out of the room and get transportation back to the airport just to try out the weather. If it was not adequate further along it would have meant returning to the airport, waiting for transportation back to the motel, re-hiring a room and waiting. In other words, if the weather did not work out, it would have meant taking an hour and a half to get from motel to airport and back plus the cost of two taxis and probably another room charge.

As it turned out my gamble on the weather paid off. The VFR corridor I could see continued all the way and the reporting stations along my route became VFR much sooner than the forecast had anticipated. I flew for several hours with heavy rain just beyond my left wing and with several miles of good VFR on the other side. If I had waited at the motel until the phone reports were good, it would have taken an extra day to get home.

Now, my "accommodations" at the airport obviously were not as comfortable as those of a good motel. But in those circumstances they were much better for me. They were ready immediately and were comfortable and adequate for a good night's sleep. I had simple but very adequate food – easily and quickly. And being anxious to get home, the extra day saved, without taking any chances in actual bad weather, was very valuable to me.

How To Ask Permission to Camp

At a *true* wilderness strip you will usually be completely alone and often in a place where camping is obviously natural and to be expected. In more civilized spots there may be some question about camping. Contrary to advice often given in guidebooks, you frequently do not have to look for an established campsite.

"Camping" means different things to different people and so it is important to phrase your request to camp properly. In this way you frequently can obtain permission to "camp" at places that are not considered true campsites. If you leap vigorously out of your plane and run up to the owner of a place and immediately ask, "Is it okay to camp here?," you may easily give him the wrong impression of boisterous revelers around a large dangerous campfire. If, instead, you slip out of the plane and go over and chat quietly for a few minutes expressing your admiration for the place and then ask, "Would it be okay to stay here tonight and sleep under the wing?," you will often get an affirmative reply. A little later add, "Would it be okay to put up a little pop-up tent next to the plane?" Almost always this will be agreed to and then you have it made! (If you are an outdoor lover, it is probably unnecessary to add that your subsequent conduct regarding noise, littering and fires can make or break the situation for the next one who comes along.)

Campfires

In many of the usual camping spots campfires have had to be banned for safety and ecological reasons. With the lightplane you can sometimes reach locations such as the rocky shores of a Canadian lake or an Alaskan river where there is an abundance of surplus driftwood and safe spots for a pleasant campfire.

CHAPTER 16

CAMPING EQUIPMENT – BUT FOR THE LIGHTPLANE

SPECIAL REQUIREMENTS FOR FLYING CAMPING • CATALOGS AND SHOPS • CAMPING CLOTHING • SLEEPING BAGS TENTS • FOOD • KITCHEN GEAR • KNIVES LIGHTS • PACKING THE PLANE • BACKPACKS

In an airplane, danger from weather comes not only while flying. Your plane can also get you into trouble because it can so comfortably and easily take you into areas where the weather can become unexpectedly very cold. Preparation for unexpected weather is much more important to the flier than to the automobile traveller. With an automobile, you can, as a last resort, run the car and its heater, or, except for very unusual conditions of snow, ice or flood, leave and drive home or to somewhere warmer. Not so by plane. It is not practical to use an airplane cabin heater on the ground. And the very weather that brought the cold may well have also brought such poor flying conditions that you are stuck where you are. So proper gear for cold and wet weather is much more vital for the plane traveller.

SPECIAL REQUIREMENTS FOR FLYING CAMPING

The popularity of camping today has created many reliable sources for camping equipment. These are all oriented to the needs of either the backpacker or else the automobile or boating camper. The requirements of the airplane camper are often quite different from either of these. These needs of the flier are not known or understood by most of the camping equipment industry. Some of their gear is wonderfully suited for the airplane but they do not know it. This chapter discusses these unique needs of the aviation camper and shows how to evaluate data about regular camping equipment for the somewhat different needs of the flier. It is by no means a complete run down on camping equipment but is intended instead to emphasize some of the usually neglected features that are important for flying.

Many of these differences concern the weight and size of various items of equipment. While fliers are always concerned about the weight, they are not concerned about it to the degree that the backpacker is. In flying, weight is usually considered in pounds, whereas for gear that is going to be personally carried, sometimes over arduous trails, *ounces* are considered important. Therefore, what you will be reading in most camping catalogs or be told by camping store salesmen will tend to be slanted to the value of extreme lightweight. For flying use, often an item just a few ounces heavier may be much more satisfactory. It is the same with size. Saving space is important in a plane, but not nearly as much as for the backpacker who must utilize every square inch of his pack. So, again, items a little larger than those that are best for backpacking may give much more comfort and utility and still be small enough for the airplane. At the other extreme are items for automobile, camper, trailer or boat camping which are much heavier, larger and more complex than the airplane camper needs or can carry.

As we shall see, there are other important differences in equipment for the flier. One of the most important of these is reliability and quality. The airplane camper is potentially much more at the mercy of his equipment than most other types of campers, and it *must* be adequate and reliable.

Keeping Warm and Dry

Pilots often consider cold weather gear only for trips to rather extreme areas or for camping, and it is an often overlooked safety measure for ordinary flights. A modern plane is so comfortable and usually so reliable that the average pilot often jumps into the plane with no more thought of foul weather preparation than if he were driving down to the drugstore. Actually, an unexpected forced landing, even on many usual and "ordinary" routes could necessitate remaining outside all night where the weather can become very cold during the night. Available warm clothing has sometimes prevented much misery and even actual tragedy to the uninjured occupants of airplanes forced down unexpectedly even though sometimes not far from civilization.

On our frequent family flights to Baja, except in the very hottest months, I have always insisted that each passenger bring a very warm jacket. Often these are not used, but I feel that they are an essential part of our survival gear because the desert often becomes very cold during the night. (Besides, if everyone doesn't have one, I, as captain, might have to give up mine. Kids, especially, will try to slip into the plane in only a tee shirt and jeans and will be very vocal in protesting at having to go and get a warm jacket. They will also be the most vocal if they are cold later on.) At the worst, the jackets can double during the flight as comfortable pillows, (but be careful that the zipper edges don't rub against the plexiglass).

Hypothermia

Hypothermia is a popular word today which means (literally), "I'm freezing to death in this cold, especially with this wet rain and wind." It describes the actual lowering of body temperature (which is the prelude to freezing to death) due to cold, especially when increased by wind or wetness. There are complicated charts showing the "chill factors" for various winds at various temperatures. The actual figures do not mean too much except to warn us how much the wind can add to the effect of low temperatures. Being wet greatly increases the loss of body heat in two ways. First the familiar evaporation which consumes heat and secondly the direct loss of body heat by conduction which is 25 times faster in water than in air. Make no mistake about it, being cold can be more than just the discomfort; it kills people every year. The answer, of course, is to be sure and have adequate clothing and, if you are camping, adequate sleeping equipment – sleeping bag and probably a tent.

Insulation (for Jackets and Sleeping Bags)

The insulation and construction of warm jackets and sleeping bags will be discussed together because the factors are the same for both. It is important to understand the properties of different insulating materials in order both to get the best value for what you spend and, more importantly, to be sure that you get equipment that will be adequate for your safety if you get into a really adverse weather situation. (If you already have good equipment or are not contemplating purchasing gear soon, you may want to pass over these technical details lightly.)

As you know, there is no magic about the value of any material for insulation – it is simply *a material's ability to keep heat from escaping from the body.* Flying puts a little different priority on some of the relative values of insulating materials as compared to backpacking or other on-the-ground type of camping.

Goose Down

Goose down is the legendary material *par excellance* for cold weather gear, but like most legends, the full story has other sides to it. Down is the lightest practical insulating material, which is its chief claim to popularity because, as we have just seen, for many outdoor uses a weight saving of ounces is important. Down also compresses greatly and reversibly. A fluffy, thick garment or sleeping bag can be stuffed into a very small sized package for carrying, yet it can be quickly removed and shaken to regain its full thickness and softness. (The first experience with a down item often consists of playing at stuffing it into the tiny sac and then taking it out again and watching it expand.)

For backpacking this compressibility is important, because every cubic inch on a backpack is important. Now, volume is also important in airplane loads, but it is not *this* important. So, some other material that does not compress *quite* so well can still be satisfactory for carrying in airplanes. Down's superior compressiblity actually works against it because when you sit or lean against something with a down jacket on, or when you lie in a down sleeping bag, the pressure easily squashes the down and it loses its insulating value in this spot. Other disadvantages of down include its greater cost than other materials of equal insulating value; it is cumbersome and difficult to clean; it easily clumps up, (thus down articles have to be made with extra construction features to keep the down in place); it dries out very slowly if it gets wet; and most critically, its insulating value is nil when it is wet. Therefore care must be taken to keep the article always dry.

Synthetic Insulators

But for use in an airplane or where you will not have to carry the article very far yourself, the superior lightness and compressibility of down are not as important and the features of other materials may do a better job. Consider seriously some of the "second generation" synthetics, man-made materials that are just a little heavier and bulkier than down for the equivalent amount of insulating value and which do not compress into *quite* as small a package. However, they do have better resistance to losing warmth by compression between the body and another object; they dry faster and clean easier; they are a little more rugged in construction; a little cheaper; and most importantly they retain some insulating value even when wet. Therefore, items made with them are still very adequate as regards size, weight and compressibility for airplane use and are better in general handling and care and in keeping one warm when wet.

Fiberfill and Polarguard

The two best known are polyesters, Fiberfill made by Dupont and Polarguard made by Celanese. Both are basically adequate, but each has minor advantages and disadvantages. Polarguard is made up of long continuous filaments which stay in place with less baffling or quilting than is necessary for the shorter 2¼ inch fibers of Fiberfill. The shorter fibers are more slippery against each other, however, and thus feel softer. Fiberfill has hollow fibers which is claimed as an advantage. Probably you should try on comparable quality garments of each and see which feels best.

Shells

Nearly of equal importance to the insulating material are the features of the cloth parts of the garment surrounding and holding the insulation. Consider here the weight of the material, its wearing quality, its waterproofness and its general feel or comfort.

Other things being equal, a lighter *weight* garment usually feels better. A thin nylon feels more comfortable than a heavy canvas, but within reasonable limits, the few ounces of weight difference per yard that may be very important for a backpacking article is insignificant for flying. The *ability to take wear* is important because garments used in plane camping may be subject to more wear than in many simpler hiking uses. Nylon's superior wear resistance can often be increased by combining it with other materials, usually a fine cotton. This will also often look better as it wears. The *waterproofness* of a material may be very important. A problem arises here because as most materials become more resistant to letting outside moisture in they at the same time become more resistant to letting inside moisture out, and perspiration builds up inside the garment. It is easy to make a completely rainproof garment, but such a garment worn for even a short time while doing any exertion at all becomes as wet inside from perspiration as it is outside from the rain. Various mechanical vent designs to let the inside moisture out without letting the outside wetness in all seem to be only partly successful at best. Thus the problem remains.

For *light exposure to rain,* the best answer seems to be some of the "water resistant" materials (usually a nylon-cotton combination, or a special expensive fine tight cotton weave). These will keep rain from soaking in for a little while and still let perspiration escape. For *heavy rain* and long exposures a completely waterproof garment is probably the only answer, but it should be worn no longer than necessary. Some people are very happy with a poncho that is waterproof, yet lets plenty of ventilation come in the open bottom. Other find that these flap and blow around in the wind too much to be tolerated. A final, and quite subjective quality, is *the "feel" of a material.* Some jackets and sleeping bags made of soft-finished material just feel better than others and will probably give the owner greater satisfaction, so long as this good feeling does not mean the sacrifice of other more important qualities.

New Concepts In Insulation

New "space age" materials are starting to be used. There is a Teflon membrane that will apparently let water pass through in one direction only. When this is bonded to layers of protecting fabric, it is said to be more waterproof than coated nylon and more permeable to water vapor than the cotton-polyester fabrics. If experience in field use is as good as the laboratory tests, this should be excellent. It is undoubtedly just the beginning and before long there will be materials with exactly the qualities we want, which will make our present materials as archaic as the heavy woolen and canvas articles of a generation ago are now.

Layering

A beginner is often tempted to get extremely warm items on the theory that "you can't be too warm." Actually, this is not so, and some of the very thick "expedition" type jackets and sleeping bags are really much too warm for comfort unless one is out in a blizzard. More experienced outdoor people usually use lighter built items but use several layers of them as necessary. This layering method gives much better flexibility for different degrees of weather. In many regions, the temperature can vary with the sun and at the end of a day of perhaps shirtsleeve weather, successive layers will go on as the temperature falls. The reverse process is carried out in the morning as the sun gradually warms things up.

Don't Let Yourself Get Chilled

When you are out on your own in the wilderness, it is very important not to let yourself become really cold. When you first *start* to feel chilly, put on another layer. It is easy to put this off if you are occupied with something else until you feel *really* cold, but if you wait until then the reheating process will take up considerable unnecessary body heating capacity that could have been saved if you got into warmer clothing sooner. (It's the same as keeping on the front side of the power curve.)

Some parts of the body are much more sensitive to cold than other parts. The head and upper torso account for some 80% of the body's heat. If you can keep this area warm, other areas such as arms and legs can be left in the cold without much adverse effect. The head is extremely important as both a sensor and radiator of body heat, yet it is often neglected when planning cold weather gear. Conservation of head heat is especially important for men as they get older and the natural insulating material gets thinner.

Catalogs And Shops

Where do you obtain these articles? There are two primary reliable ways, either through mail order catalogs or directly in "trail shops." Many catalogs are a gold mine of information about materials, construction methods and specific models. After getting a basic knowledge from several catalogs, one can then go to a trail shop and look at various items much more intelligently and with much less dependence on the whims of the salespeople.[9]

Avoid Cheapies

A word of caution is in order about some of the lightweight jackets and sleeping bags often advertised in newspaper ads or found in discount-type stores. Often called "survival jackets" and sometimes described as "jungle tested" (whatever that may mean!), they are often questionably made items with very thin actual insulation. They may feel warm and comfortable when tried on in the heated store or at home, but in real camping conditions they can be a disaster. Your actual survival *could* depend on the quality of your equipment and you should be sure about how and of what it is made.

CAMPING CLOTHING

What you wear and take on a camping trip will depend on your own tastes and desires, but some suggestions are given here that may help you in choosing what you will be most comfortable with. Always remember that extra warm clothes can't do any harm, and can be invaluable if you are in an isolated camp and the weather becomes unexpectedly cold.

Pants, Shirts, Blouses, Skirts & Dresses

Take and wear what you are most comfortable with and what you feel is most appropriate for the type of place that you will be staying in. Remember that wool is warmer than cotton and this difference is much greater if they are wet. Blue jeans are notoriously poor insulators and they have contributed to the demise of many hikers and climbers who have become lost or injured in freezing weather. Yet their popularity is well deserved because of their comfort and toughness, and if backed up with a reserve set of insulated long underwear, especially of wool or synthetic, they should be safe. In any area where wet weather of any duration may be encountered, each person should always take along one complete covering of wool clothing. *Remember that "wet weather" does not mean that it has to rain.* Wet clothes will result also from snow already on the ground or from nighttime atmospheric condensation, or "dew," that occurs very frequently in otherwise excellent weather. An invaluable homemade item is a pair of ordinary blue jeans with patches of waterproof coated nylon sewed on the entire seat area and over both knees. With these, damp ground or sand can be sat on, a tent can be crawled in and out of in damp grass without the persistent clammy wet feel that such activities often result in with non-altered jeans. But there is still the lightness and "breathability" which would not be present in pants made entirely of waterproof material. Shorts are often most comfortable for warm areas, and sometimes "walking shorts" can be used instead of underpants. Then long pants or a warm skirt put on or taken off as the temperature varies.

Jackets and Sweaters

As noted above, several relatively light layers are more flexible than one heavy layer. A good jacket will almost always have a flap that snaps closed over the zipper. This helps keep cold from entering at the thin zippered area and also provides insurance against that occasional but desperate zipper failure in the wilds. An uninsulated "mountain parka" of water repellant cloth replete with multiple pockets is an indispensable article. Alone, it is often just right for comfort in mildly cool situations, and over other layers of warm things it gives an added boost to comfort in really cold places. Of course at all times the pockets are invaluable. If you are using down filled garments, there should also be at least one heavy wool sweater or one synthetically insulated jacket for wet situations.

Underwear

This will vary between what is normally worn for "ordinary" climates and insulated items for colder areas. Stress physiologists may tell us how their investigations show that arms and legs are relatively unimportant in relation to body warmth: but the practical fact remains that a very cold situation can be much more comfortable and relaxed with a pair of insulated "longies." Many types and brands are available and the choice is quite straightforward. Women find that long nylon stockings or pantyhose are a help when it's cold, but in more extreme situations, insulated underwear may be needed.

Insulated tops (or one piece combinations which are warmer than separates) are often used, but it is awkward to shed an underwear top, especially for the ladies, and it is easier to use whatever is normally worn and put on an outer layer such as an additional sweater, which is simple to take off and put back on as the temperature varies.

Hats

Hats are often treated as casual accessories, but they are actually very important because the head controls a large part of the body's heating mechanism. Have a thick knit cap that will come down over the ears. It can be wool or synthetic, an inexpensive Navy watch cap or one of the colorful and attractive ski caps. Visor type caps that shade the eyes often add to comfort. For wet weather there should be either a waterproof hood on a jacket or parka or a waterproof hat. Vanity is unimportant in wilderness rain, and a large sized plastic baggie over the wool cap will do fine.

Gloves

Mittens are warmer but gloves are more flexible if the fingers have to be used. For ice fishing and other cold weather outdoor activities a small catalytic hand warmer that burns lighter fuel is popular. These require separate fuel, a funnel and a cover and are probably more complicated to carry than they are worth for an airplane trip.

Footwear

For flying and ordinary walking use whatever is usually preferred. Hiking boots are too cumbersome for some pilots to use for flying but are popular with many for hiking. The many varieties suggests that there is no one that is ideal. Selection may be simplified by understanding that most hiking boots are designed to give a platform for backpacking and the weight and stiffness of the boot should be proprotional to the amount that will be carried. Camping fliers will probably not be backpacking very much, so a lighter hiking boot should be satisfactory. A synthetic lugged sole called "Vibram" seems to be the most popular and is good for general hiking and rock climbing (but watch out on small wet logs and large round wet rocks – they slide sideways very easily). Vibram comes in a bewildering variety of compositions but you just have to trust the boot manufacturer to have used the proper one.

A hiking boot gives protection to the ankle. Once while hiking up the side of a volcano in Guatemala, wearing only low shoes, I slipped and twisted my ankle; nothing really serious, but disabling enough to wipe out the remainder of the trip. I am sure that if I had been wearing a higher hiking boot there would have been no injury.

Waterproofness may be very important in footwear, and for this there does not seem to be any completely satisfactory solution. Many hikers are happy with boots made of the full thickness of the hide (i.e., not "split grain" which is lighter and softer – and more comfortable) to which coats of waterproof dressing have been applied. For protection against tall wet grass, low bushes or snow, they seem happy with "gaiters" of waterproof nylon or plastic which run above the boot to the knee. I have not found this combination very satisfactory. The full grain boots are quite heavy and stiff and the waterproofness, especially in wet brush is sometimes inadequate. The best thing I have yet found is a pair of very inexpensive 12" high rubber boots which slip off very easily for flying and on easily after landing. They are not quite as comfortable for long hiking as leather boots, but in morning dampness or in streams or marshy areas they are completely waterproof.

Socks

Usually under hiking boots or rubber boots two layers of sox are worn: a thin one next to the foot made of wool (or a combination of wool and a synthetic for better wear), although some strongly favor silk. Over this is a thick long sock for warmth and a cushion. Usually this is predominantly wool but some favor Olefin or other materials which claim to have a "wick" effect and better take the perspiration away from the skin. It would seem as if this combination would be too warm for ordinary weather but after a little getting used to it is quite comfortable. (In Canada heavy socks are sold by weight and a variety of thicknesses is available.)

Nightclothes

Long ago, an experienced woodsman told me that one should remove as much daytime clothing as possible before going to bed at night. Winston Churchill even used to put on pajamas for his daytime siesta; but, of course, he had a valet. For most people sleep is more relaxing when they are undressed for bed. When camping, this raises the problem of overnight storage to protect daytime clothes from the dew which can often make them damp, clammy and uninviting in the morning. Prevent this by stuffing them into the stuff sack the sleeping bag just came out of. If they are damp, it may be better to put them in the sleeping bag with you. They will feel a little cold and clammy at first, but this feeling quickly disappears and your body heat will help dry them out overnight.

It will probably take some experimenting to find out what is most comfortable to wear in your sleeping bag at night. I have found that in most all climates I am most comfortable with an ordinary cotton grey sweat suit top, the kind with the fuzzy flannel feel. I like one with a hood and I avoid the cold metal zippers. Whatever you choose, a shift from daytime clothes will usually give you more relaxation and better sleep. In really cold conditions it may be better to sleep in your clothes (without shoes, naturally). This will keep you warmer, and you don't have to get cold while putting on clothes in the morning.

SLEEPING BAGS

After clothes, the next most important item in your camping equipment is a sleeping bag. Again, it's down with its lightness and compressibility versus the synthetics with their extra strength and superior performance when wet. Combined bags with down tops and synthetic bottoms are also available. Again, beware of the cheapies that are adequate only for slumber parties or a Boy Scout overnight in the local park and are potentially lethal in really severe conditions.

It is tempting to get a thick bag rated to far below zero, but such a bag is often too warm in ordinary weather, even with the bottom unzipped, and a lighter one may be more comfortable. In the occasional extreme conditions you can always put on more clothes and get the effect of a thicker bag. Some folks use a thinner bag and get an extra insulated liner so that the sleeping bag can be "tuned" to the weather. You should also consider your personal metabolism as well as the conditions you are likely to be in. Some naturally "warmblooded" people need less insulation than more chilly types. If you are the hardy "under the stars" type, you will need a warmer and more waterproof bag than if you will be sleeping in a tent, the plane or other shelter. Sleeping bags are often described by the number of seasons they are supposed to be adequate for. Thus a "three season" (i.e., Spring, Summer and Fall) bag is not as thickly insulated as a "four season" bag. You can determine relative warmth capacities a little more accurately by considering the weight of the insulating material in bags of the same overall size. Catalogs give details of various types of construction which can make a difference, too, but if you buy from a reliable supplier you can count on his having used the proper construction method for his bag.

Again, for airplane camping, requirements are a little different than for backpacking. The so-called mummy bag is the warmest for its weight because it stays close to you all around, but it is more confining and less comfortable than the "modified mummy" or even the full "rectangular" style that lets you move around more fully and rest better. The slightly increased size and weight of these latter are not significant for airplane use.

Most bags have a two-way zipper so that the bottom can be zipped part way open and give added ventilation and less warmth in warmer conditions. When this is done, always keep the two zipper sliders at least 10-12 inches apart to avoid too much strain on a shorter unzipped section.

Storage

Proper storage of an unused sleeping bag is very important. A bag, especially a down or short fiber synthetic one, is a very delicate article. When not in active use it should not be kept compressed in its small "stuff sack," nor should it be folded over a coat hanger or hung by one end. This puts an abnormal strain on the delicate stays and baffles inside the bag which keep the insulation in place and maintains its insulating capacity. A bag should always be stored in a large sleeping bag storage sack (or even a large plastic trash bag with a few small vent holes) so that there are no stresses on it and the insulating material can fluff out and not remain compressed for a long period of time. The traditional "cool dry place" is best, but at least try to avoid extremely damp or hot storage places.

Liners

Sleeping bag liners that are detachable will take the soil and can be removed and washed when you get home. This saves the entire rest of the bag from many washings or dry cleanings which, at best, are not good for it. Be sure and get a liner of a material that feels comfortable to you. They come either with a smooth nylon type material or a thicker flannel feeling cloth. The majority of experienced campers do not use liners, however. They feel that they tend to bunch up or pull uncomfortably when turning over, etc., and they don't believe that the protection for the sleeping bag is significant.

TENTS

A modern lightweight tent can give a big increase in the comfort of your expeditions. A tent can protect you from wind, moisture and greatly increase the warming capacity of your sleeping bag and clothing. Many of us still think of tents in terms of the heavy canvas monsters of the Army and do not realize the tremendous advances made by modern tent makers. Today's tents vary greatly in cost and complexity.

Makeshift Tents

The simplest tent is a piece of waterproof material thrown over a wing with the sides secured by rocks. (Low wings need less material for this and are blown around a lot less; but watch your head if you suddenly sit up at night.) Similar in simplicity is the "tube tent," a nine foot long tube of

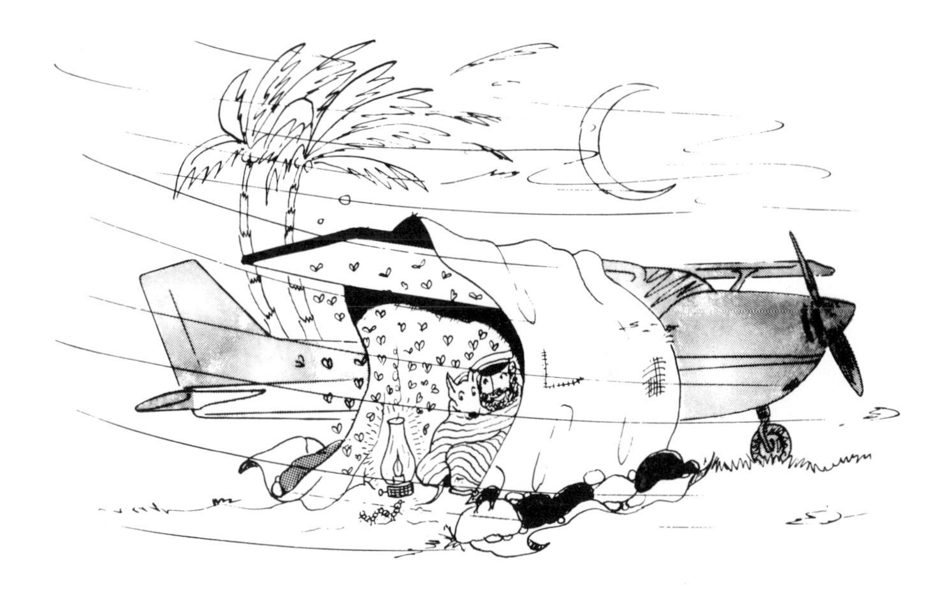

light (3 mil) plastic usually six feet in diameter. A thin rope is run through it and secured to suitable supports. Rocks or other heavy things are placed inside it so that a triangular cross section three feet each way is produced. Campers then crawl into the triangular tube. Plastic sheets, tarps and ponchos are also used in various configurations in similar fashion. All of these give an increase in shelter and comfort but all have the major drawbacks of ends that are open to bugs and wind. If these openings are closed the result is no ventilation which can cause discomfort from the build-up of humidity inside.

True Tents

There are two general categories of real tents. The simplest and cheapest are made entirely of waterproof material. Being waterproof these are also proof against the escape of the inside moisture that builds up in the tent from perspiration (even in cool weather), and an intolerably humid situation can result. If there are enough openings that are adequately sheltered from possible rain and screened from bugs, the result can be quite comfortable – if you don't mind keeping the openings uncovered. A more sophisticated (and more expensive) type of construction uses waterproof material on the bottom parts of the tent and builds the uppers out of porous, breathable (and non-waterproof) material. A "fly" of waterproof material is then mounted so that it covers the entire top area but leaves several inches of "breathing space" between it and the tent wall. Thus there is protection against rain or condensing dew; and even with the tent windows and doors covered for privacy or to avoid wind or light, there is still adequate ventilation.

A third type of tent is being experimentally marketed made out of the "one way" membranes of Teflon covered with protective cloth. So far heavier and more expensive than the other types, it is a promising experiment.

Various tents have various ingenious methods of *suspension;* combinations of lightweight aluminum poles, stakes, guy ropes and shock cords. I have a very satisfactory one that has a complete frame of lightweight metal rods which is quick to assemble and needs no stakes or ropes. It assembles in less than five minutes and after being erected can be moved around. This is very handy if the wind changes and starts to make a smokehouse out of your tent, or for moving it periodically to keep it in the shade of the wing as the sun moves. Stakes are provided for use in heavy winds, but I usually just put two heavy things in each downwind corner and it stays in place very well.

Delicate Floors

All the lightweight tents have very thin floors which are delicate and not always 100% waterproof. Always try to clean out all rocks, branches and other sharp irregular objects from where the tent will be pitched. Also help boost the floor's waterproofing by being sure the spot has good drainage, and never put one over a depression that may become a puddle in the rain. If you can carry it, a piece of heavy waterproof material placed under the tent floor will protect the delicate floor from rocks, etc., and increase its water resistance. Be sure that the edges of this pad do not extend beyond the sides of the tent. The first time I tried this was with a large piece of old canvas tent given to me by an Indian trader in northern British Columbia. Since a large section was available, we made it extend out beyond the tent a foot on all four sides. The first night of heavy rain, I found out that this had been a mistake. I awoke thinking my tent had become a bathtub. Water seemed to be coming through the floor and would not run back out again. It turned out that the waterproof canvas sheet extending out beyond the tent edges was catching the water running down the sides of the tent and collecting it in a slight depression under the tent. The waterproofing of the thin nylon floor was not equal to keeping the puddle of water from seeping through under the pressure of the tent occupants. But there was no pressure to force it out again, so it collected inside the tent. So, if you use an under-tent pad (and it's a good idea to protect the thin floor from rocks and other sharp things underneath), be sure that the pad is slightly smaller than the tent bottom and does not extend beyond any edge of the tent. Even so some water will run down around and holes should be cut in strategic places of the pad to let this escape.

Tent Fever

The size of the tent has some psychological considerations. Too small a tent can give an effect of semi-claustrophobia, especially in prolonged bad weather when you are in the tent for long periods. Of course, the very large tent that would be best in this respect would be too large and heavy to take; but at least avoid the very small ones. For one or two people a "three man" tent should be the minimum size, to allow some moving around space and room to take extra things into the tent. Again, you are not going to carry it on your back and a little additional size and weight will more than pay off for you.

Flame Resistant Fabrics

If you will be smoking or cooking near your tent, especially if it may be pitched near the plane, be sure that the material is flame resistant. (Actually, most camping veterans seem to be leery of cooking in a tent and prefer to either arrange a shelter to cook outside the tent or else eat cold food. Keeping food out of the tent is also a safety factor if bears are around.) The so called "CPAI-84 specifications" provide for flame resistance. If purchased in California, Massachusetts, Louisiana or Michigan this is a legal requirement, elsewhere you will have to check on it to be sure. A flame resistant material will burn as long as it is in contact with a flame but will not continue burning by itself. The possibility of a blazing tent under the wing of an airplane does not need any further elaboration for any pilot!

Make-Your-Own Kits

Kits for making your own insulated jackets, sleeping bags and tents seem to be quite popular. A kit comes with all the large cloth pieces already cut to size and if you can run a sewing machine you can save and still get a better made item than if it was factory produced – so goes the sales pitch. Well maybe! But if your time is worth anything at all, the savings are questionable. The quality of the article will depend, of course, on your expertise in making it, but remember that you will be working with very thin lightweight materials and that factories have specialized equipment for this. A lot of kit making is quite monotonous, such as cutting and sewing a great number of identical small connecting pieces, etc. I have had a lot of experience in making quite sophisticated electronic kits, usually very easily and successfully, but I never felt so lost and bewildered by instructions as I did with the one tent kit I attempted. My wife rescued me and took it over. She is an experienced seamstress but she, too, found it very confusing and difficult. If you are interested, look the kits over; some of them are very intriguing, but proceed with caution.

Air Mattresses and Foam Pads

That grand old lady of Baja California, Bertie de Meling, one of the pioneers who founded the wonderful Meling Ranch, told me when she was around 84 years old that she knew that she was getting old, because ever since she was eighty, she had had to use a sleeping bag when she went on one of her frequent overnight horseback trips into the nearby rugged and wintery San Pedro Matir Mountains. This made me feel like a real city softie because even in much more comfortable camping locations I always take along a full length 1½ inch thick foam pad and, if there is room, a second one. I also am very fond of a piece of sanded and varnished 3/8 inch plywood, 44 inches by 20 inches which flattens and smooths the ground underneath the pads from shoulders to hips. If no one is looking, I also often stuff a full sized foam pillow into a small sack and sneak that along too. But I am admittedly psychoneurotic on this point and know that I do better with comfortable sleep.

But for most everyone who does not rough it often or long enough to really get used to the bare ground, a cushion is really necessary. Even experienced woodsmen, these days, usually carry at least a shortie foam pad.

The old standby was the *inflatable air mattress.* These are still popular and used by many. They deflate and save space. However, *foam pads* now come with thin durable waterproof nylon covers and can be rolled into a relatively small roll. They are more reliable and probably more comfortable than an air mattress. Foam also does a much better job of insulation under the sleeping bag where the weight of the sleeper makes the insulation thin. They come full sized or half (shoulder to hip) size. For the average airplane camper who is used to a comfortable bed at home or in a motel, these are a very much appreciated addition to his "roughing-it gear." There is now a *"self inflating air mattress"* which combines the best of both types. A foam pad is surrounded by an airtight nylon shell. Open the valve and the foam expands and "breathes in air." Blow in a little bit more and close the valve, and you have a combination air mattress and foam pad. In the morning open the valve and slowly roll it into a small roll as the air goes out again.

FOOD

Food – important for enjoyment, maintaining energy and even occasionally survival. Some people enjoy creating full course gourmet meals in the wilderness, while others forego gastronomic pleasures and concentrate on the wondrous beauty of nature while chewing on beef jerky and munching granola. Most of us fall in between these extremes. I always take things that I like, but that are easy to prepare (usually with a can opener), and that are still nourishing enough to assure the extra energy that may be needed in the rough.

Here again, advice that may be excellent for backpacking must be interpolated for the different situation that flying creates. Flying's slightly less stringent weight requirements allow a better quantity and quality of food. An on-the-ground camper usually has to take his whole food supply with him, whereas, the flyer often can stop along the way for resupplying, and meals can be boosted with a box of fried chicken, fresh fruits, vegetables and dairy products. (A small insulated cooler is valuable for this. The soft ones that collapse when empty usually pack better than the hard foam ones, but be sure it doesn't leak on other gear.) Backpackers often skimp on food. Some even plan on using their spartan fare as a weight reducing aid. A pilot, on the other hand, must be sure that the diet is adequate to keep him in top shape for flying requirements. Any food that causes gas may be a minor problem on the ground, but can make serious discomfort in the air.

Dehydrated and Freeze-Dried Foods

Various methods can remove the water from foods and make them much lighter and easier to store for long periods. It is claimed that the camper can thus enjoy all kinds of tasty foods almost instantly and almost without effort. Others, less enthusiastic, are kind enough to say that the flavor gets better the longer you are away from home cooking. Actually, some are excellent and some are blah, and part of the planning for a trip might well be testing sample meals at home where the refrigerator is handy to supplement the failures. There are various sized packages, single servings or enough for two or four servings and larger cans which are much cheaper if you can use that much. When trying them at home, note the amount that each package will make. Outdoor appetites are hearty, and the packages are usually already optimistic in the number of servings they claim to make.

If you are in areas where there is lots of good water, dehydrated foods are more practical than in desert areas where you have to carry your own water. It does not make sense to pay the higher premium for dehydrated food, if you have to carry the water to reconstitute it anyway. On the other hand, if there is ample water where you are going, the dry food can save a lot of space and weight. (Dehydrated foods make good survival rations too.)

Dehydrated Alcohol

For campers who like a cocktail before dinner, weight can be saved by taking 95% alcohol which eliminates over half of the water found in ordinary liquors. This is available in Alaska and a few other states under the tradename "Everclear," and in most liquor stores in Mexico labeled simply "Alcohol." In some areas if the liquor laws and liquor tax regulations permit, it can be obtained from a cooperative pharmacist. It is 190 proof and must be diluted accordingly, but when equal parts of water are added the result is identical to vodka. This alcohol is distilled from grain rather than potatoes as vodka is, but all alcohol molecules are identical and when refined to this degree, no trace of its heritage is left.

There are several drawbacks. In many places it is very difficult if not impossible to obtain; many people do not like vodka and want the traditional liquor that they are used to; and finally there is sometimes a psychological turnoff. The fact that alcohol is actually a drug is often shrouded by the social acceptance of the various packaging it is dressed in. The "bouquet" of a fine wine or the "smoothness" of an expensive scotch or bourbon somehow shield one from the harsh reality of the drug they contain. When faced with drinking pure "alcohol," the sudden realization that it is actually a drug is disturbing to many people.

Canned Food

I usually depend primarily on canned food, the 8 ounce "buffet" size for one or two people and larger, more economical sizes for more eaters. In many cases, the taste quality is better with canned foods: at least it is a taste that most of us are used to. In low water situations, try to drink all the juice from the canned foods. You flew it in and it will save on precious drinking water.

Wild Foods

The food supply can sometimes be boosted by such things as freshly caught fish and certain wild plants which can be very tasty as well as genuinely "organic." These things should be considered an added bonus and never depended upon, however.

Check Out The Supermarket

Before you load up on the relatively expensive dried camping foods, wander thoughtfully through your local supermarket. You will find many articles that are excellent for camping and at much less cost than the camping food packages. Some items, such as quick rice or dry soup mix, are nearly identical to the camping foods yet much less expensive. Other possible items are the old standby Spam or small tins of boned chicken or turkey which provide good quality protein concentrated in convenient packages. Use your imagination and you will find many suitable things that will appeal to you.

Planning The Food Supply

Some thought and planning is essential for your trip's food supplies because, unlike regular shopping, you must be sure that you haven't omitted an essential item – you can't run back to the store if you start to make a meal and find that some vital ingredient was forgotten. It is essential to make out a rough outline for each of a day's meals and then add everything up to make a shopping list. A very simplified example could go like this:

Breakfast	Lunch	Dinner
Can of fruit	Bread or crackers	Can of meat or chicken
Pancake mix or dried eggs	Cheese or peanut butter & jam	Canned vegetables (1 or 2)
	Dried soup mix	Can of fruit or dried fruit
Coffee, tea or cocoa	Candy	Cookies
Sugar	Coffee or cocoa	Coffee or dried milk

Then add up the total number of servings of each item for a day for one person:

Canned or dried fruit 2
Canned vegetables . . . . .1 (or 2)
Canned meat or chicken (can substitute baked beans, freeze-dried products, etc.) 1
Dried eggs or Pancake mix. 1
Bread or crackers. 1
Cheese, or peanut butter and jam 1
Dried Soup Mix. 1
Cookies 1
Tea, coffee, cocoa or dried milk 3
Sugar and salt . . . . .as needed

Figure out the size and number of cans or packages needed, depending on the size of the group, and then multiply these figures by the maximum number of days that you will be eating this way. (It is best to plan sizes of cans so that the entire can will be used up each meal. It's hard to carry around loose wet food in a plane, and refrigeration is often a problem.) This gives you a basic shopping list from which you can make substitutions and changes for variety while keeping the total amounts the same.

The method just described may seem very simple and obvious, but you must figure out in this way, or some other accurate manner, a *definite* total list of what you will need. Any more casual shopping method such as, "One of those, and those, and those – Hey! these look good, let's take along a couple of them," will result in badly needed items missing at a time when they can't be replaced, or else the carrying along of a lot of heavy stuff that you won't use.

Shopping En Route

Whether you take it all at the start, or plan to resupply along the way will depend on many things, such as the size of the party, number of days involved, the weight-carrying capacity, etc. which only you can figure out. If you can safely carry the weight, it's generally best to take as much as you can from the start. It is often hard to get to a store on a wilderness trip. If you do find one it may not have the things you want or the size of cans you need, and prices usually will be considerably higher in out-of-the-way places. If you do plan to resupply along the way, take along a copy of the shopping list, although experience may have dictated by then a revision in amounts or varieties. Whenever you do run across a food store, fresh fruits, vegetables and dairy products can often freshen up your canned menu.

Some folks like to put all items for each meal together in a plastic bag; others like to pick out each item and arrange meals the way they feel at the time.

Goldbench, Alaska. *It's a long way to the grocery store from here. (In addition to true survival food, it's good to have some "emergency" food for when you are overdue for a meal and away from your source of supply. Take along a box of breakfast bars and a six pack of 6 oz. self-opening cans of fruit juice. One of each of these in your pocket will provide a tasty and well balanced emergency meal.)*

Emergency Food

For real wilderness trips there should always be several day's emergency food in addition to the regular supply. In the sparcely settled areas of Canada and in Alaska the amount is spelled out by law (7) (8). A small stuff sack filled with freeze dried packages is a good way. In an emergency, local water can always be sterilized and used (10) . Some old timers take dry dog food for their emergency supply. It's light in weight, keeps well, is nutritious even for humans, and it is not likely to be eaten unless there is an emergency need for it.

If children, and especially hungry teenagers, will be along, the food situation should be discussed with them in advance of the trip. Be sure that there will be enough for adolescent appetites, and also be sure that everyone understands and agrees to the importance of the food supply and the danger from casual snacking on the planned amounts.

KITCHEN GEAR

What you will need for kitchen gear will depend on how large your party is and how much actual cooking you plan to do. Always try to keep it as simple and light as possible.

Stoves

You must have at least a simple stove if only for heat and hot drinks in emergency chilly situations. A minimal stove that I like is put out under several brand names and is a simple small burner that plugs into a shaving cream-type butane can. It quickly disassembles into a small disc for easy carrying. This will boil one pan at a time very adequately. An extra fuel can should be taken, and the spare marked so that you will run one dry before starting to use up any of the other. For more extensive cooking a larger stove may be wanted. They come in all sizes and types and are best chosen from camping catalogs or at trail shops. As usual, for flying, certain features are more important than for other types of camping. Many use either gasoline, (not Avgas, unfortunately), kerosene or alcohol. These require "loose" fuel and the sudden changes in altitude in flying can cause storage problems that would not be present for on-the-ground use. Butane or propane in cans or cartridges is more expensive but easier to carry in a plane. It does not work well in very cold temperatures. Look them all over, but keep in mind your unique needs for flying.

Cooking and Eating Utensils

Many varieties exist and the choice depends upon your habits and how much cooking you will be doing. Try to keep them as few and simple as possible. In the wilds you can take many short cuts in table setting, food preparation and serving. Dishwashing is sometimes primitive and difficult, so try to make multiple use of pans, spoons, etc. and use as few as possible. A pan used for making soup, for example, can be used without washing to boil water for most freeze dried foods. A nesting set of pots that fit inside each other will save a lot of space. Ones like the Swiss SIGG with rounded bottoms (called "Billies") are easier to clean. Carry them in a sock for protection because they are easily dented into uselessness.

Knives

Each person should have a rugged sharp knife. Novices tend to carry large, impressive looking sheath knives; more experienced woodsmen usually carry smaller ones. Folding knives are much more comfortable for wearing on your belt while flying than the larger non-folding sheath knives. If you get in the habit of carrying your knife always on your belt, you will have it with you when you need it. Otherwise, you may not take the time to look for it and put it on for small land excursions, which is often just when you need it most. A good quality pocket jack knife is fine if you are happier carrying that, and won't lose it from your pocket. There are many good looking but poor quality knives on the market, so be sure that you get one that will hold a sharp edge since it can be very important to you. If you will be hunting and your knife may be used for skinning game or for any other situation where it will be used a lot, a small sharpening stone will make its use easier. These also come with belt cases so as to be handy when needed.

LIGHTS

There are many possibilities for light at night depending upon your preferences and habits. If you can get into the routine of going to bed early, you can use the sun much of the time. Others may want to play cards all night and will need a long lasting substantial light. Geography will play a part here too. Mid-summer in Alaska needs very little supplemental light; more equatorial regions, especially in the winter, get dark earlier and there is more need for light.

Flashlights

For those who retire early or who may read only a little at night, a two cell flashlight will be adequate. The D cell size flashlight weighs about twice as much as the more compact C cell size, but the batteries cost the same and have twice the capacity. Flashlights smaller than C size have very short light duration. Flashlight batteries will last much longer if they are used the way their name implies – intermittently. Frequent short rests will prolong battery life considerably. You can even select the bulb for maximum brightness (PR-2) or maximum duration with slightly less brilliance (PR-6). A spare bulb must be taken. Some like a square or lantern type battery light, and while these are heavier, they do not need to be held all the time.

Gasoline Lanterns

A pressurized gasoline lantern gives an economical and reliable light for a long time. They are a little large and cumbersome for flying and unfortunately won't burn airplane gas, so separate fuel is necessary. If you do take one, the mantles survive transportation much better if the lantern is carried upside down. Some people save the original carton and varnish it to make a durable carrying case.

Other Lights

Other types of portable lights include *propane cylinder mantle* lamps, battery flourescent lights and others that are good if they will fit into your size and weight capabilities. The *plane's cabin lights,* if any, can be used *sparingly,* but this starts up the electric gyros. (I have often wished for an FAA approved separate master for lights only.) Of course, the gyros run all the time when you are flying, so perhaps this extra use is insignificant. It's just that without the noise of the engine they sound as if they were working so hard. There are *portable flourescent 12 volt lights* that plug into the cigarette lighter and give much more light per ampere than the usual incandescent bulbs. These are light weight and popular with many campers. See them at trailer supply stores.

Emergency Light

No matter what else you have for lights, you will need a flashlight and a waterproof one is best. Camping stores carry very high quality waterproof, lightweight, aluminum ones. Get in the habit of putting the flashlight every night in *one* convenient place (such as a side pocket of the tent) where you will *know where it is in the dark.* In a sudden emergency at night (someone is ill, or your dog has located a skunk), it helps to have a quick light without having to rummage through who-knows-what in the dark.

Paper Towels

This is a seemingly minor item but it is listed because it is extremely useful for an unbelievable variety of uses. For checking the oil level; cleaning up a multitude of spills, slops and messes; as napkins and table mats; packing material; emergency toilet paper – keep a large roll handy. They are worth their weight in gold.

PACKING THE PLANE

Okay – you have selected and collected all this gear, now how do you pack it in the plane? As usual, there are so many variables and personal preferences that only very general suggestions can be made. You will find that you will be constantly revising your methods and ideas. Even after you have had much experience, new innovations will occur to you. The number of people along and the size of your plane will dictate some limits. More people take up more space in the plane, of course, but also require that more equipment be taken. As a very general rule, for any extensive camping, two seats per passenger will probably be required. Other important variables are the length of the trip and the possible opportunities for restocking the supplies.

Taking out the back seat in many planes will greatly increase the cargo space and save the 20 to 25 pounds that the seat weighs. *(But watch that weight and balance extra carefully if seats are removed!)* This extra space can mean that many items can have a more or less permanent place, and this is one of the secrets for an enjoyable trip. Nothing is more frustrating than having to paw through stacks of belongings to find something that is urgently needed. This is often a problem no matter how well you plan and arrange things in the plane, and the most valuable tip in this whole book is to emphasize *having a definite place for everything and trying to keep everything always in its place.* Needed items seem to vanish just when you need them the most, and nothing puts more of a damper on the relaxation or enjoyment of a trip than the temporary loss of some item that is vital at that moment.

A System

Here is a system that works very well for me, but you will probably have to modify it for your habits, preferences, equipment and space. In general, emergency gear, surplus supplies, and other items not frequently needed should be on the bottom or furthest from access, wherever it may be on your plane. Your sleeping bag, mattress roll(s) and tent, if any, form one unit which will be needed all at the same time, and usually only at night. If they can be together in one part of the storage area, it will be easy to find them all at the same time, and they can be bypassed as a unit when you are getting to other things.

For food, several strong medium sized *cardboard cartons* are used (approximately 8"x 10"x 14"), one for each meal's supplies. Lately I have been experimenting quite successfully with shallow lightweight *"cat litter" trays* of different colors. One carton (or tray) has all the breakfast makings in it, another for lunch and another with dinner materials. Cans of food are laid down so that labels are easily read. (If put on end, each can has to be picked up and turned sideways to see what is in it.) If you will be getting fresh supplies occasionally, you will want an *insulated container* for perishable foods. Soft plastic ones have the advantage of being collapsible to save space when not completely full, but watch out that they don't leak on other things. Another carton for "hardware" contains other frequently used items such as a can opener, roll of paper towels, flashlight in the daytime, insect repellant, etc.

Some people like to use *nylon stuff sacks* of different colors for different things. These are good for camping type clothes, etc., but for cans of food or other small articles, it is hard to look over the contents as they all slide together in the bottom of the sack. Grocery store sacks should be avoided for the same reason and also because they wear out very quickly and dump their contents from the tears and holes that appear. If you will need any "dress up" clothes or have any that wrinkles will effect, you may want to have a suitcase or two to give these better protection. Most camping type clothes do very well in stuff sacks or duffel bags, and these are easier to pack in the plane. It is helpful to start out with a *written list* of what things are where; it can save a lot of searching through various containers at first. And again, always try to replace everything in its proper container. Under the temporary pressure of a moment when something else is waiting to be done, it will often seem like a lot of extra work to do this, but in the long run, it will make much less work and will make the whole set up much more smoothly running and relaxing.

BACKPACKS AND DAY PACKS

A confusing multitude of knapsacks are divided into two general categories – larger "Backpacks" which are built around a wood or metal frame and can carry considerable loads, and smaller "Day Packs" usually without frames.

The Backpack

The modern *backpack* is probably the most efficient machine ever devised for carrying heavy loads on the human back. The frame is arranged to keep most of the weight off the shoulders and upper back and transfer it to a padded belt around the hips. Large loads can thus be carried with relatively little energy because the bones of the hip girdle carry the weight rather than the muscles of the back. For carrying heavy loads, it can't be beat. For the backpacker, it is also handy because it has numerous compartments and zippered pockets so that it acts as a chest of drawers; a "home away from home", and is the central core of a backpack trip.

In an airplane, quite to the contrary! A backpack is a miserable beast in an airplane. The frame catches around other gear constantly, way back in the storage area where it is very hard to free up. It is bulky and takes up the cargo space of several boxes or stuff sacs but gives only a fraction of this much storage space. The compartments and pockets which are so handy for the backpacker are nearly useless in an airplane because they are so hard to get to. So, unless you are going to be carrying gear away from your plane for an overnight bivouac, do not take a backpack. Seldom have we included categorical advice, but this is a justified exception.

The Day Pack

The *day pack*, on the contrary, is a must! Get a comfortable one, large enough to carry your camera, an insulated jacket, perhaps a thin rain poncho and still have space for groceries that you may buy, or rocks and other specimens that you may collect. The right size is 2500 to 3000 cubic inches. This will free you from a camera around your neck, a jacket tied around your waist and makes the carrying of groceries and other items incredibly easier than any other way. Again categorical advice: take a day pack, and you will wonder how you ever got along without one. You may start using it at home for shopping trips and other non-camping activities.

CHAPTER 17

STRIKE IT RICH

BE A FLYING PROSPECTOR

Throw a gold pan in with your other gear and become a prospector! Gold pans are light, inexpensive and fun and easy to use. Poking around looking for promising places and panning for gold is a hobby that is easy to get really turned onto. If you are lucky, you may find enough to pay for your gas, and in many places there is still the real possibility of that big strike!

Psychologically, panning for gold is alot like fishing. It gives an "excuse" for being outside in beautiful country. Our Western civilization has given many of us a compulsion for needing to be doing something or for having some "reason" for just sitting and enjoying nature. Fishing and gold panning both fill this need without much effort or concentration. Finding a little gold gives the same psychological lift as catching a good fish.

There are potentially worthwhile gold prospecting areas spread over large parts of Canada and Alaska, some areas in many of the western states, and a few spots in Mexico. Many of these can be reached by the private plane and some of the most promising areas are easily reached only in this way. It was the search for gold that opened up many of the wilderness areas that we can now enjoy. The old prospectors and later the commercial miners developed many of the roads, towns and isolated airstrips in these wild areas. As the gold fever subsided, many of the developments regressed and the wilderness took over again, but the roads and airstrips sometimes remain to the benefit of the wilderness seeker.

When the earth was new, gold started out being mixed with all the other elements. Later, it was often melted by volcanic activity and deposited with the other things in the lava. Erosion removed the rocky elements in many places, but the heavier gold tended to stay behind. A lot of it was washed down by streams and rivers, and as it went along, its weight made it sink to the bottom through the loose rock and gravel to be deposited in crevasses in the "bed rock," which is a solid layer of rock that keeps the gold from penetrating it. It was deposited more heavily in places where the river's current slowed down, and the heavy gold dragged behind and was left on the bottom. Such slowing down usually occurred on the inside of sharp bends in the river or where the river widens and the current slows. Success in gold mining is related to the ability to understand these really simple physical principles and figure out where the gold might possibly have been deposited.

Gold Panning

Gold pans traditionally were made of steel. Now, black plastic ones are lighter, work equally well, are rustless, and show up the gold better against the black bottom. Gold panning is really very simple. The basic principle is that gold is heavier than the sand and gravel it is found in. Scoop up about a quarter of a pan of gravel or sand from a likely place in a stream along with about a half a pan of water. The idea is to slowly rotate the pan so that the heavier gold sinks to the bottom while the gravel or sand is gradually and gently sloshed out over the edges. You may need to add more water very carefully to complete the removal of the sand or gravel. When there is only a little left, swirl it around gently with the fingers of your other hand and, if you are lucky, a definite yellow color may be seen in the bottom of the pan. If you do find any, you will need small containers to keep it in. To show it off at home, put it in water in glass vials. The water magnifies it and makes it look larger. So take along a gold pan. When you get in the right place you may get very enthusiastic about mining. If you become really interested, you can get a light weight gasoline powered pump and a small sluice box and operate much more efficiently. The smallest of these was designed for backpacking and weighs about 35 pounds. Another lightweight method for locating larger nuggets of gold uses a sophisticated metal detector. The complete story is fully and interestingly told in a book by Matt Thornton.(11)

CHAPTER 18

SAVE DOUGH – CAMP AS YOU GO

Anyone who flies cannot be too concerned about economizing, because flying is expensive no matter how you do it. Still economizing is one of the real advantages of sleeping by the plane. Today, the increased costs of ground transportation and overnight accommodations soon equal the cost of even the best of camping gear. After that, it is all savings. And on an extended trip this can be significant.

The Relative Savings

With today's inflation, anything giving actual prices can quickly become obsolete, but relative values can still be outlined. It seems as if inflation is increasing the prices of accommodations and other travel costs more than it is increasing the costs of camping equipment, probably due to the relative amounts of labor involved, so that the relative savings by camping is becoming greater as inflation progresses.

A very rough estimate of camping equipment costs might be:

Sleeping bag	$100
Insulated jacket and other clothes (although some of these would be needed for wilderness trips without camping)	$150
Tent	$100
Foam pads, etc.	$ 20
Stove and utensils	$ 30
Miscellaneous	$ 20
Total	$420

If these items are used for two weeks the equivalent cost is $30 per day, adding $5.00 for groceries, makes it $35.00: *compare this to daily non-camping costs of lodgings at $20.00; meals at $11.00; taxis, etc., at $4.00 – which totals $35.00 per day. In two weeks the camping equipment is paid for, using this comparison, and after that the savings are almost infinite.*

Family Camping

Camping has extra advantages is children are along on the trip. Children take naturally to camping, they like the outdoors, the campfire and the freedom. Motels and hotels can be confining to children and have restrictions that can be forgotten when camping. The "patter of tiny feet" can be disturbing to a downstairs guest in a hotel, but no one will even notice happy feet running along a camp trail.

The economic advantage of camping is even greater with children. Extra hotel rooms for children multiply costs, and satisfying ravenous young appetites at restaurant prices can soon cripple many budgets, whereas camping meals cost no more than eating at home. By using camping, many flying families have been able to have family fun together and to see many interesting and educational areas that could not have been afforded otherwise.

Often it is best to camp on only part of a trip. Some places may not be good camping spots or it may be helpful to break up the camping routine occasionally by a stop at a place with hot showers and perhaps for some "meals out" for a change from camping food. But, even so, the part of the trip that uses camping will afford good savings.

Economical Equipment

Various items of camping equipment for flying are described in Chapter 16. The costs of this equipment can vary a lot depending upon where it is purchased. As we have seen, all items should be of top quality. Occasionally you may be grounded by bad weather and really dependent on your equipment so it must be adequate for any weather that may come along. If the price is low, be sure that it is a legitimate "sale" of a good quality item and not a thin item made for selling at the low price.

Quality items are sold at a variety of price levels. There are stores that specialize in outfitting wealthy sportsmen. Their outdoor clothing is often extra well tailored and of better appearance than more ordinary camping clothes and this may fill the need of some flying campers. Their other equipment is usually the same as the top-of-the-line items at other stores but sometimes with a premium added for their label. Most ordinary camping stores, or "trail shops," carry standard brands and the prices for the same quality items are usually rather standard. There are a number of "discount" type outlets which supply good quality equipment at somewhat less than the normal retail sources. One of the most popular of these is the REI, Inc. a co-op which was started in 1938 and now has over half a million members. Regular brands are sold at minimum costs and excellent quality items are put out under their own label at considerable savings. Members also get a substantial dividend on purchases during the year. Their mail order catalog is an excellent source of camping information and there are retail stores in several West Coast cities. (12)

Used Items

Often you can pick up good bargains in used camping equipment. Usually, however, the items will not be really suitable for airplane use. If they are, be sure that they will be adequate for the critical situations that you may encounter.

An unexpected source for some camping clothing, such as wool pants and sweaters, is the salvage stores run by various charitable organizations. If you don't want to pay for new items for only occasional camping use, this may be for you. You will probably be surprised at the good quality of often only slightly used clothing that can be obtained for only a few dollars. Camping clothes should be made of wool and today's synthetics resemble it so much that it may be hard to tell what a used garment is made of. The prices are so low that it is worth the gamble to buy a likely appearing item and have your dry cleaner determine for you if it is really wool.

Disposable Hospital Supplies

Most hospitals today use many disposable items. It is cheaper to throw them away and buy a new one than to pay for the labor to have non-disposable items cleaned, sterilized and repackaged. Many of these items are very useful for camping, ranging from plastic water containers to large plastic sheets. If you know anyone who works at a hospital, perhaps you can get ahold of some of these items.

Food

It is generally advisable to take as much of your camping food from home as your weight capacity will allow. You already know where to get the best buys at home, but even more importantly, out of the way places on a trip may not have the kinds of food or the sizes of containers that you want. Then too, it is often troublesome and very time consuming to locate food stores near an airport on any flying trip, and especially so in more isolated areas.

Time Savings

Just as it does in so many other ways, flying can save much time on camping trips. It is often hard to put an actual dollar value on the time saved, but, for most of us, time does have value and using the lightplane to save time also saves money. (This is an often used way, and usually a legitimate way to rationalize the high costs of flying.)

Time savings can be especially significant in wilderness camping trips, because the areas visited often take a long time to reach on the ground. They can be reached in much less time by plane and for the time available for any vacation, more different areas can often be visited.

Last year I flew to Bella Coola, a very isolated little town halfway up the coast of British Columbia, (probably the most scenic spot I have ever been to). Flying time in my 182 was 17.3 hours. Driving time to it would have been approximately 45 hours. Or, if one was passing through the "jumping off place" for Bella Coola, at Williams Lake and wanted to swing over and see it, it would take 1.5 hours by plane compared to 11 hours of very hard driving on land.

Everyone has to calculate for himself what such time savings are worth, but it is obvious that the potential savings are significant and that in the wilderness the time savings are greater than that for "regular flying" because of the poorer land travel conditions.

CHAPTER 19

GROUND TRANSPORTATION

BICYCLE • MO-PEDS • PORTABLE BOATS

As most pilots have found out, a major drawback to private flying is the occasional lack of mobility on the ground. The airplane has moved quickly and easily to the destination but then, after arrival, the efficiency of the transportation comes to a screeching halt. Pilots often find that they get a lot of exercise walking. If there is no phone nearby or if a taxi or other ground transportation is not quickly available, the family may start saying, "See, we should have taken the car after all, because that would have given us transportation now." However, if the trip was to the usual paved strip in civilized country, transportation can usually be obtained somehow.

When landing at a wilderness strip this problem may be greatly intensified. A fortunate choice of location may result in recreation facilities handy to the strip, and this is one of the criteria listed for consideration in selecting a strip. If properly (and luckily) done there may be a river, lake or camping area right next to the strip so that ground transportation is not needed or even wanted. But there are still many strips with very desirable features that are some distance from the strip – too far to walk to, or at least too far to carry much camping gear to. Some other strips with very good close-by features may have additional things well worth visiting that are not near the strip.

This problem has resulted in the creation of numerous contraptions designed to fit inside the plane and provide a means of overland travel after landing. Usually, these are some variation of folding minibike or a take-apart motorcycle, but other types of collapsible motorized creations are sometimes seen. Many of these can do a fair job of providing transportation on the ground, but they all have drawbacks as airplane cargo. Motorcycles are quite heavy and take up a lot of cabin room. The time and effort of assembling greasy parts means that it will not probably be used unless there is really something worthwhile to see or unless some time is going to be spent at one location. Folding minibikes are easier to set up, but they are usually small, uncomfortable and underpowered. Other contraptions have similar drawbacks: If they are large and powerful enough to be practical on the road, they are usually too large and heavy for storage in the plane. Licensing and acceptance by local authorities, to say nothing of liability insurance, may also be problems.

Bicycles

On one trip, I took my ten speed bicycle. With the quick mount front wheel and the pedals removed, it packed in the plane without too much hassle. I saw some country that I couldn't have without it, but it turned out to be mainly a demonstration of how fast an airplane covers the countryside. During a landing it seems as if the plane is only crawling and interesting looking things are seen that seem to be "just down the road" from the airport. The bicycle soon makes it obvious that the 80 MPH or so landing approach speed is really fast, and the spots which seemed so nearby from the plane are reached only after long and hard pedalling by bicycle. A cold headwind also affects a bicycle much more noticeably than it affects a plane.

Mo-Peds

The best solution that I have found so far to this problem is the "Mo-Ped." This is the little half bicycle, half motorcycle that is becoming popular in many urban areas. Weighing about 65 pounds, it is relatively easy to get in and out of a plane and will cruise at 20-25 MPH with very low gas consumption. If the going is uphill or rough, it can be pedalled to give the 49 cc motor a boost. Comfort is equivalent to a bicycle. It's lightweight makes it surprisingly responsive and it is really fun to ride. It is surprisingly good off the road on trails and paths. It is street legal most everywhere and in many states doesn't need licensing. Mine is an Italian make that has proved its reliability and durability through years of use in Europe. An increasing number of brands are becoming available so make the selection remembering that short overall length and light weight are important. For getting in and out of an airplane without damage, avoid the ones that have mufflers or other tender parts directly on the underside.

With the back seat out of my 182, it is really easy to get it in and out of the plane (a six foot length of a one-by-four board helps too). I often take the pedals and tail light off for a little smoother loading and unloading and sometimes use it with these still off. It runs on any grade of Avgas with 2% oil mixed with it. I keep two cycle oil in a plastic baby bottle which seals well and has measurements on the translucent side. This in an oil proof box together with a quart plastic bottle for mixing, take up only a small space and are light weight. It is still not really ideal airplane cargo, but it is by far the best thing I have yet found to solve the ground transportation problem.

On a recent airplane camping trip through northwestern Canada and much of Alaska, I put over 300 miles on my mo-ped and I got out to see places and things that would be really impossible in any other way. It is also a conversation piece and an ice-breaker because it is a happy looking little machine and most everyone wants to look it over and ask about it.

If your dog is along, it will have a great time running along beside you, but two cautions are necessary. The engine makes just enough noise that a vehicle approaching from behind may not be heard and, as I found, a city dog's feet quickly become blistered almost to the point of temporary total incapacity. So either work the dog up gradually or else get an astringent solution to apply beforehand. I am told that little "hiking boots" are also available for dogs! (21)

Portable Boats

If the cargo allowance will permit, this ground transportation concept can easily be extended to water. Two man inflatable boats are available which weigh about 20 pounds and when deflated are the size of a middle-sized suitcase. A small 1½ horsepower outboard motor weighs around 25 pounds.

The result is not exactly a water ski boat, but it will enable you to get out on isolated lakes where the fishing can be really fantastic. Boating capability can add another dimension to your wilderness exploring. From an already isolated airstrip on the edge of a wilderness lake, a boat can take you into spots that are almost literally untouched by man.

On this Canadian trip with my mo-ped, I discovered several very secluded cabins on a beautiful and otherwise completely deserted lake.(13) I quickly got to dreaming of how nice it would be to have one as an ideal fishing retreat. Land up there must be very inexpensive and most of the building materials for a log cabin are growing right on the spot. A little more thought brought the sober realities of the impracticalities of such a venture. Yet, with the mo-ped and by taking along a small inflatable boat with a small outboard motor, it would be easy to set up a camp on such a lakeshore which, while not equal to the comforts of a cabin, would achieve perhaps 85% of the enjoyment of one and without the various responsibilities that the year round care of a distant cabin would have.

CHAPTER 20

EMERGENCY REPAIRS AND TOOLS

When something appears to be wrong with the plane at the usual flying spots in civilization it is usually easy to get it looked at by competent mechanics, the problem diagnosed and, if necessary, fixed. The pilots responsibility is only to see that the work is done. But, when something seems wrong in the wilderness far from any professional mechanical help it is a different story. The pilot is now faced with the responsibility of determining both if the plane is still safely flyable and, if it is not, with seeing if he can repair it himself.

The FAA regulations are very specific as to what can be done by a pilot to his own plane. FAR 91.163 and FAR 43 spell this out *in detail.* Normally, a pilot or plane owner is allowed to do only "preventive maintenance" which means "simple or minor preservation operations and the replacement of small standard parts not involving complex assembly operations." It then goes on to list 25 specific operations that are allowed, and are presumably the only things that are allowed. These are very simple things such as changing the oil and spark plugs or repairing seat upholstery. Anything more complicated than these items must be done by "certified maintenance personnel."

What if you are out alone in the boondocks and something more complicated needs repairing? The regulations do not seem to allow for this contingency but it would seem that if property or people could be shown to be in danger that it would be permissible to make repairs and fly at least as far as the nearest authorized person to inspect the repair.

Insurance companies are also interested in your repairs. If repairs are done in violation of FAR's it may invalidate insurance claims, although to what extent this would be carried out is hard to determine in advance. My own insurance agent assured me that in any location where authorized mechanics are not available it is okay to do any necessary repairs that you can and have them checked out as soon as practical. It seems to be a complicated, indefinite subject and anyone wanting an authoritative answer should consult an attorney who is versed in aviation matters.

From a practical standpoint, there are times when it seems sensible to go ahead and make necessary repairs even though they are not authorized. If something is loose and can be easily tightened back up with a screwdriver or wrench, it seems dumb not to do it even if the part is a "no no". Once I was landing at a small isolated strip when I had complete electrical failure. After checking out the battery and its connections (permissible) I located the trouble. A wire had fallen off its terminal on the alternator (a non-permissible item to fix). The alternatives were the illegal procedure of replacing the wire and its connector or the legal procedure of leaving it alone, propping the plane and flying without radios or other electrical equipment to perhaps several airports until an authorized mechanic could be located. Needless to say, I replaced the wire and it went fine. Mexico requires that a Mexican mechanic do any repairs in that country. It's not for safety but in keeping with their overall policy of only Mexicans working in Mexico. Mexican mechanics are usually very good, although they are more used to bailing wire techniques rather than using new parts. It is partly to save costs and partly because parts are hard to get in Mexico.

Minimal Tool Kit

In the wilderness, as well as probably everywhere else, a minimal kit should be taken. There have been many cases where disasters have occurred because of the lack of a simple tool. Several years ago a nearly new Cherokee Six had to make a forced landing on a beach in Baja. A bracket was loose on the engine which could have been easily fixed with a simple screwdriver, but none was available and the incoming tide soon totalled the otherwise intact aircraft.

I always fly with three screwdrivers, a large and smaller regular one plus a medium sized Phillip head. Also there is a small crescent wrench and a small (1/8") socket wrench set with several sockets of the most common sizes. These weigh only a couple of pounds and are kept in a small bright colored thick nylon sack. I try to discipline myself to keep them in the sack and the sack in my flight bag. Tools seem to be disposable items around our house and if the airplane tools are used repairing skateboards and motorcycles, they may not be available when really needed for the plane.

CHAPTER 21

ANIMALS. THOSE YOU FIND AND THOSE YOU BRING

WILD ANIMALS • WHALES • FISH • BIRDS
PROTECTION AGAINST WILD ANIMALS
VENOMOUS CREATURES • TAKING PETS INTO THE WILDERNESS • DOGGY DIPLOMACY

Experiences with animals in the wilderness are two kinds. *Wild ones* that are found there and *pets* that are brought along.

WILD ANIMALS

Animals have always been one of the chief attractions of the wilderness and are one of the things that draw many people to it. Their popularity used to be largely for hunting, but as animals become more scarce and there is less need to use them for food, hunting has declined somewhat. Now animals are often only observed, photographed and enjoyed. Nearly everyone is excited by seeing any type of wild animal anywhere, and finding animals in their natural state in the wilderness can be one of the really high spots of a trip. A wild animal somehow awakens something from our prehistoric past and gives a seasoning to a trip in the wilds. Seeing wild animals gives almost the same type of satisfaction as catching fish (and the tales afterwards often grow the same way). Television producers find it worthwhile to devote much time to films of wild animals. In the National Parks, the bus rides at Mount McKinley that go into areas where animals can be seen are booked solidly well in advance, while at Yellowstone the removal of the bears from the tourist areas has caused bitter complaints from many.

An almost endless variety of animals can be seen in the wilderness: reading about an area in advance of visiting it will indicate the kinds of animals that can be anticipated there. The more that can be learned, ahead of time, about the identification and the habits of the wild creatures in any area, the more interesting it will be to discover and to watch them. When looked for from the air, animals will rarely be seen in the summer when the leaves are on the trees. Much of Alaska, for example, is still loaded with wild game, but it is all hidden from air observers in the dense underbrush and trees. As soon as the leaves go in the fall, however, the animals suddenly appear nearly everywhere.

WHALES

Usually, the larger land mammals are thought of as being the interesting animals, but with a little study many other creatures will be found perhaps even more interesting. Observing the migratory activities of the great California gray whale is an absorbing hobby for many western flyers. The giant animals can easily be seen from the air as they migrate up and down the coast, usually swimming near the surface and close to shore. A flight to Baja California during the winter can give an amazing opportunity to observe them from the air as they breed and give birth in the shallow lagoons. (See page 235.)

photo by Phil Grignon

A Whale in Scammon's Lagoon. *My friend used a 300mm telephoto lens while I flew as quietly as possible. We found that if flown over at much under 1000 feet, whales would submerge. Incidentally, photography from Cessnas (and probably other high wing planes) is simplified if the side window hold down bracket is unattached. The window then will open completely and stays up by itself while in flight.*

FISH

Fishing is often the reason for a flying trip to wilderness areas. Unpaved airstrips can give a private pilot access to a vast number of places where there is still fabulous fishing. These include all types of fishing: stream, lakeshore, lake boating, surf or "deep sea" fishing. Sometimes fish too, are just looked at, either by snorkeling, scuba diving or even through the bottom of glass-bottomed boats. (14)

BIRDS

Just a little study of birds can add greatly to their interest. One can easily learn to watch for eagles, hawks, buzzards, osprey as well as all the varieties of smaller birds. In the more tropical areas, many wild parrots and other very colorful birds are abundant and easily seen. (See Pg. 236.)

Possible Mid-air With A Bird

In less civilized areas where birds are more abundant, there is more chance of a mid-air with a bird. The chances are still very slight, but it does happen. Even large planes can be felled by tiny birds, viz. the jet crash on take off at Boston some years back when its engines were clogged by a large flock of small starlings. A lightplane engine is not very vulnerable to a bird strike; its vital components are pretty well covered up, although presumably a bird could cover and clog the air intake. (In such an event, carb heat on might provide a clear air intake). In rare cases a windshield can be broken by contact with a bird in flight. Even if no injuries result, the noise and sudden rush of air would be a frightening experience. In settled country a landing at the nearest safe place to leave the plane for windshield replacement is indicated. In the wilderness, landing as soon as safely possible is also the indicated procedure, but the new windshield may be a problem here. Repairs should be undertaken very cautiously because makeshift repairs will be subject to hurricane-force winds even in slow flight. If *absolutely necessary,* the plane can perhaps be flown slowly with the broken windshield to the nearest available help, but remove all passengers and cargo (except for survival gear), wear warm clothes and a wool cap and be prepared for a stormy flight.

Wild Kingdom

Camping in isolated areas is often done in the center of the realm of many wild animals, more animals than are sometimes immediately apparent. For wild animals have a great ability to locate man and to keep out of his sight. The visitor will be seen by many more animals than he will see.

The danger of wild animals can be very real, but it is often exaggerated. Wild animals are not nearly as aggressive toward man as is often thought. If it possibly can, almost every wild animal will avoid man. Even the traditionally fierce wild wolf will try to run away from man every time. Even if cornered or caught in a trap, a wolf will cower before a man brandishing a stick. Accurate research has failed to substantiate the legends of wolves chasing men. A wolf pack may watch a campfire at night, and the howling may sound menacing, but there has never been a substantiated report of an attack on man. (15)

Bad Actors

Almost all other North American animals are similarly more afraid of man than man is of them. One rare exception is the polar bear which has been known to sneak up and attack seal hunters on ice floes, but there are no airstrips even remotely near any polar bears today.

There are two animals that, under certain circumstances, can be dangerous for wilderness fliers, the bear and the moose. The grizzly bear is more temperamental and aggressive than other bears, although they are said to be unable to climb trees. But any bear can be dangerous. If given advance notice of man's presence, a bear will almost always leave, usually quietly and without any indication that he was near. However, if startled by being come upon suddenly and without warning, or if any situation develops that looks to a mother bear as if her cubs are in danger, a touchy situation can develop. Since bears are noted for unpredictability, either of these situations can be the trigger for a sudden unexpected attack. And despite his appearance of slothful clumsiness, a bear can move fast – much faster than any man possibly can. Again unpredictable, his attack may carry right to a human, or may stop suddenly at any point short of this. Defense in this situation is difficult to say the least. If you come upon a cute cuddly bear cub "all alone" in the woods, leave it alone, do a quick 180 and a full power retreat! For he is being watched carefully by a hidden babysitter who is very possessive, has a short temper, and can outrun you three to one, outweigh you five to one, and is carrying twenty double edged switch blades which she knows how to use with deadly efficiency!

In addition to the maternal instinct, hunger will also sometimes make a bear shed its usual tendency to avoid humans. When busy gathering berries in the woods, or when concentrating on catching salmon in rivers, bears are well known for ignoring nearby humans. Animal researchers have spent many days literally surrounded by fishing bears without any significant problems. (16)
After all, the bear has no natural enemies in the wild state, and if he has no previous unhappy experience with man, he may see no reason to take time to be concerned with man when there are more important things for him to be doing. In a wilderness flying trip, bears might occasionally be encountered while they are gathering berries or possibly honey, but there are no airstrips near places where bears fish for salmon, (and I have really looked for one)!

Pine Lake, Y. T. (17) *These bear tracks appeared in a muddy spot very near the plane in the middle of the day while we were away from it for several hours. There was no other evidence of a bear's presence.*

PROTECTION AGAINST WILD ANIMALS

Often there is no way to tell for sure if there are bears in an area. Reading up on an area ahead of time can give a general idea of where bears are, but there is no way to be sure about any specific place. Almost any wild place that has vegetation is a likely spot for a bear to visit; and bears are sometimes found quite near to civilized areas. The existence of farms and grazing cattle is no guarantee that bears may not be around. If it is some distance from a settlement, a garbage dump is often a very popular place for scrounging bears. Bears seem to like spots with quite dense underbrush high enough to hide them, and it is in these places where you are most likely to come upon one unexpectedly. In areas like this or areas where the terrain is irregular enough to afford hiding places for bears, it is best to avoid sudden confrontation by making noise when walking along. Do this by singing, yelling, or wearing a small "bear bell" or other noisemaker. If a dog is along, it will often know when a bear is nearby and start barking or seem very restless. (Nothing makes the fur on my dog's back stand up as much as when she sniffs a fresh bear dropping.) It is wise to restrain dogs if bears are expected. A dog can usually keep away from a bear's reach, but it may run back to its owner bringing its adversary with it.

Besides being startled or having its cubs threatened, about the only time a bear will bother a man is to get after food. When camping, keep all food sealed up so that the odor will not attract the always hungry bear. Wash your hands and face, especially if you have a beard, after eating or touching food and before going to sleep to remove all traces of food or its odor. If possible, do not eat in your tent or sleeping bag or keep any midnight snacks in them. Tragedies have occurred when bears go after food in occupied tents or sleeping bags. Bears may also consider other things such as soap or toothpaste good to eat.[18] Do not keep food in your plane in the wilds unless it is well sealed up. In his attempt to get at the half candy bar in the side pocket or the part of a sandwich that slipped down under the seat, a bear can easily rip through the light skin of a plane and quickly make it extremely unairworthy.

Guns

So the best protection is avoidance. In a situation where avoidance may not be possible, a gun may be a necessary safeguard. Experienced woodsmen have strong and conflicting opinions as to the best gun for this. Some favor a heavy calibered revolver, (certainly the easiest to carry in the woods), others scoff at this and insist that a rifle is the only thing to have (although the proper caliber is the subject of other heated discussions). Others feel that a properly loaded shotgun is best. To cut through all the emotionally charged claims and figure it out scientifically, consult the tables of muzzle energy for different cartridges given, among other places, in the Remington Arms Catalog.[19] In these, even the most powerful hand guns come out a weak third. The larger rifles are good for longer distances, but for close-action defense, a shotgun is probably best. The Alaska Fish and Game Department, which presumably should be knowledgable about this, has issued to their men who are unavoidably exposed to bears, 12 gauge pump shotguns, (supposedly a little more reliable in a pinch than the otherwise faster semi-automatics). These use a progressive loading, starting with two rifled slugs, then a 0 shot shell followed by two 00 shot shells. The theory is that the slugs will carry farthest, and if these fail, the shot will then be more effective at close range. Needless to say, this is a hairy situation at best. The bear may stop short at any time in his charge and change his mind, and presumably you don't want to have to shoot him. Yet how long can you delay while 800 pounds of furry rage is rushing at you at 44 feet per second? There will not be time for much thinking but only for instinct. Hopefully this instinct will not be needed.

If a bear is coming at you and you don't have a gun, there is not much you can do. You can't possibly outrun him; climbing a tree is no help (he can climb much better and faster than you can), possibly dodging around large trees or rocks may extend the chase a little longer. However, he is adapted to the woods much better than you are and will always prevail in the end. A time-honored possibility is to lie down and play dead, but often a bear will be so worked up that he will attack you anyway. My own theory, (which fortunately I have not had to test out), is that if his getting to you is inevitable, it would be best to turn and fight back. Pick up a stick, try to look as menacing as possible, and tell the bear to cool it. In many other wildlife encounters the weaker animal can often make the stronger give up by a show of force. If you have watched a larger dog going after a smaller one, you have probably noticed that if the smaller one runs he may get bitten, but if he stands his ground and acts agressive, the pursuing dog will usually stop the attack. It doesn't always work out this way, but if it's the only thing left to do, it would be worth trying. The best protection is still avoidance.

Moose

A moose can sometimes be as dangerous as a bear. It does not have the claws, teeth or the hugging power of a bear, but it is fast, fearless and can kick with unbelievable agility and power. A healthy moose can nearly always repel an attack by a whole pack of hungry wolves.(20) A moose too, will usually avoid, or at least ignore a man, but some, especially when with young, are irritable and have been known to attack man. The young son of a friend of mine once escaped an angry moose only by climbing up a tree.

VENOMOUS CREATURES

About the only other really dangerous creatures found in the wilds are poisonous snakes and sometimes a poisonous spider or scorpion. The danger of these has been exaggerated for rarely do they cause death; but they occasionally do cause severe injury. Rattlesnakes are thought of as killers, but death is relatively unusual from the bite of a poisonous snake. The more usual danger after the bite of any venomous animal (snake, spider or scorpion) is a severe reaction in a bitten arm or leg that can result in marked mutilation and a crippling deformity.

Like man, snakes like good warm weather, and will be found lying in the sun, especially in rocky areas. If it gets too hot, they retreat to cooler places; and if it gets too cold, they become sluggish and stay at home. So it's good weather that makes the right conditions for man and snake to meet.

Again, the best precaution is avoidance. Snakes will leave if they know a man is approaching, but if startled, stepped on, or cornered unexpectedly, they may attack. Long heavy pants and high leather boots will give a lot of protection. Small "snake bite kits" are popular and valuable. The Cutter snake bite kit or one similarly patterned should be part of your first aid gear, but carried with you in the wilds, not left with the first aid kit by the plane. Treatment requires promptness! A blade is provided in the kit and if used within the first 30 minutes (along with suction) it can be very helpful. Follow the directions in the kit exactly. If used excessively or improperly, it can do more damage than the bite. Remember to make incisions *only* over the fang marks and no more than 1/4 inch long and 1/8 inch deep. Use the tourniquet lightly, only enough to impede the superficial veins. Antivenom "serum" is sometimes needed but requires professional supervision. This means that a poisonous bite in the wilderness is extra dangerous, and care to avoid getting bitten is extra important here. The best treatment in the field is to keep the victim lying down and the bitten area cold, if possible. Ice is best, but lacking that even a cold stream is good. Aspirin for pain and any tranquilizers you may have along may help. Alcohol used to be popular as a snakebite remedy, but medically its only value is as a tranquilizer.

Poisonous Spiders

Poisonous spiders are typically found in woodpiles, on the walls of old log cabins and other undisturbed places where woodsy debris collects. The wearing of gloves and long sleeves in these situations is helpful because the spider is often not seen. The black widow spider is the famous one, but it is only one of a number of poisonous insects and is less lethal than many other less well known ones. These bites cause severe muscle spasms. It starts with slight pain at the bite with some blanching and swelling there. Rapidly it progresses to body muscle pains and cramps with restlessness. Recovery begins in 12 to 24 hours. The black widow often lives in the rural outhouses, which will still be found in many wilder places. The resulting bites are sometimes more embarrassing than dangerous.

Sting Rays

In many areas of the world, sting rays doze in shallow water on sandy beach bottoms. Their tails have a venomous barb that can give a very unpleasant sting. Like most other wild creatures, they will always leave if they know you are coming. If you help let them know in advance of your approach by shuffling your feet frequently when wading at an infested beach, there is seldom any problem.

TAKING PETS INTO THE WILDERNESS

Whether your pets will join you or not on a wilderness trip depends upon what kind of pets you have, their (and your) disposition, how much they weigh and how much cargo capacity you have.

Cats

Cats, typically, do not take naturally to travel in cars or planes, but most cats can be easily trained to enjoy riding, and many cat owners have a fine time taking them along. They are quiet, small and light and do not need much food. But they are characteristically place oriented rather than people oriented and are thus more easily lost. Also, if they get the travel bug, they sometimes tend to slip into strange cars, which is a frequent way cats become lost.

Dogs

Dogs seem more naturally suited for wilderness trips, because they love walks in the woods, will stay with you, and if properly trained, will follow your instructions. Most dogs take to flying naturally and easily. My own dog started as a small puppy. She was so content on her first trial flight that I tried a few steep turns and stalls. Her green hair should have warned me that she needed an air sick bag! But she recovered and hops happily into the plane whenever she can, drapes herself comfortably into whatever irregular space there may be left for her, and sleeps through most flights. She is bothered by turbulence, but no more so than the rest of the family. She has about 250 hours and just assumes that all dogs fly.

Avoiding Airsickness in Pets

If an animal gets airsick easily, it will help to be sure that it gets neither food nor water for eight hours before takeoff. Dogs for whom this is still a problem, or those who are disturbingly restless in a plane, can be given a safe sedative. (21) Be sure that an animal will not cause an unexpected disturbance at a time when the plane needs full concentration.

Mine has been known to suddenly start barking at cows on final; from the air, they look like other dogs to her. A dog changing its location in a plane during landing can suddenly upset the trim and airspeed by a significant amount.

Border Crossing With Pets

For taking a pet from one country to another, some travel information publications list rather awesome requirements. In practice, all that is needed for taking a dog (or cat) into Mexico is – nothing. Into Canada or the Bahamas requires an up to date certificate of rabies vaccination which can be obtained from any veterinarian or public rabies clinic. To return an animal back into the United States, the same certificate is needed and that is all. The inspection of animals at borders varies just as it does for other things in the plane. Sometimes animals are completely ignored, at other times the certificate is scrutinized long and carefully. Otherwise no problem. Central American and West Indian countries are often more strict and sometimes quarantine incoming animals for a month.

Some places in Mexico, and elsewhere in Latin America, control stray animals by periodically setting poison bait around a village. The

DOGS HUNTING DEER

Take Notice That by Authority of the Wildlife Act

ANY DOG

Found Running at Large and Harassing Deer

WILL BE DESTROYED

From DECEMBER 17, 1976 **To** April 17 1977

In the following area: REGION 4

DIRECTOR,
FISH AND WILDLIFE BRANCH,
DEPARTMENT OF RECREATION AND CONSERVATION.

2M-866-7080

Canadians can take dog control very seriously too.

local folks know about this and keep their animals at home at this time;

but a stranger can arrive and not find out about it until too late, so a dog should be kept restrained when first arriving. A leash on a dog is a good policy at any airport. Even if you have good voice control of your animal, it makes a good impression on airport officials who can help smooth your stop. It is also legally required in many places.

Chita on Matanuska Glacier, *less than a mile from the Glacier Park airstrip.* (See page 240.)

A dog can add greatly to the enjoyment of the wilderness. Conchita was born in Mexico and raised in Southern California, so she had never experienced snow or ice before. Here, she is bitterly complaining about the temperature of her feet. She is a Mexican Fish Hound, a breed not too well known in the United States that make excellent airplane and camping dogs. (For the scientifically minded the species is Fishoundus Mexicanus ridiculous.*)*

Don't Handcuff Your Dog

Never tie your dog outside at rural Mexican strips. Local dogs often feel very possessive about their territory, and a tied leash may make it impossible for a dog to adequately defend itself or flee. I know of several dogs who were thus killed by Mexican "camp dogs" when they could have otherwise easily outrun or overpowered the poorer conditioned local dogs. So don't "tie a dog's hands" in strange places. Most dogs can be trained to stay quietly inside the plane, even for a long time. My dog often spends the whole night in the plane, because when she has a restless night, she barks at things periodically all night. This might be alright if she stayed outside, but if loose, she insists on sleeping in the tent and each barking episode starts with two sudden loud barks while she is still inside the tent on her way out. In rainy weather she always comes into the tent and *then* shakes. I have yet to find an answer to this.

DOGGY DIPLOMACY

Knowledge of dogs' sexual etiquette can be a big help in a visiting dog situation. It is very rare for oppositely sexed dogs to fight seriously. Two females may manifest considerable jealous hostility, but it is two males that usually have the real fights. A visiting female dog, therefore, has an easier time socially with local dogs than a male visitor. In some isolated villages, especially in sled dog country, natives do not welcome an outside male dog because they are anxious not to dilute the blood lines of their own strain of pack of sled dogs (although it often looks as if some outside genes might help). Watching dogs' tails can also give help in judging what may happen when a visiting dog arrives. Groups of dogs form a hierarchy of dominance, or a "pecking order." The most dominant have their tails carried very high; the least dominant ones' tails are between their legs, and the others are ranged in between.[15] If two unacquainted dogs meet, especially if they are of the same sex, this order of dominance has to be determined, by a dogfight if necessary. If, on the first meeting, one's tail goes up and the other's down there usually will be no show of outward animosity. It's when both tails are up that there may be trouble.

Pet Food for Travel

Food taken along for pets will depend on what the animal is used to eating and what the weight capacity of the plane is. Canned pet food contains much water and is heavier per unit of nutrition than dry preparations (which are usually equally nutritious). Since animals can find water and can safely drink in most places, the dry food can save weight in the plane. In a real desert location where the water supply brought along will have to be shared with the pet, this advantage of weight saving is gone. If there is to be camping away from the plane, most dogs can be taught to wear a small pack (practice with it at home first), and carry their own food (and

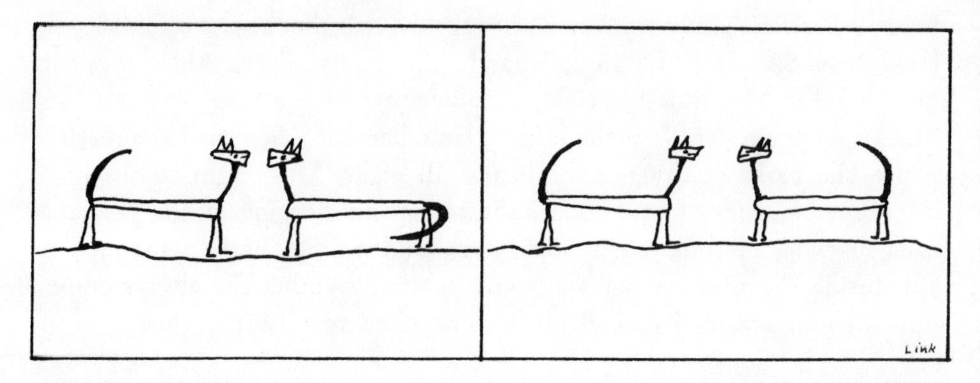

No trouble here, dog on left is dominant and both dogs acknowlege it.

Look out here! Both dogs want to be boss. Potential dog fight especially if both are males.

perhaps some of yours too). For this, a good quality of dry dog food is essential.

Some experienced wilderness fliers, even if no animals are taken along, always carry a quantity of dry dog food as an emergency ration. They figure this is less likely to be used up for snacks and will be available when needed for an emergency.

A dog's feet may need care if it suddenly starts walking and running on a wilderness trip much more than it has been doing at home.

Poisonous Salmon

At certain stages spawning salmon are easily caught by animals. Dogs are susceptible to certain parasites in salmon and certain trout, which can be fatal to the animal. So be sure your dog doesn't get ahold of any of these. (22)

Tag Your Dog

Be sure that your dog always carries adequate identification when travelling with you. Dogs that normally always stay at your side may wander away under the stimulus of new places or new dog friends. I almost found this out the hard way when my dog was taken into protective custody while she was checking out Canadian rabbit hunting at the far end of a little Canadian airport. Fortunately, the dog catcher was a pilot and he stopped by my plane to see if she might be mine. He had not booked her yet, so it was easy to get her back. After that, I made up a temporary tag from an airline baggage tag. Now, she has a permanent metal tag that says "please call collect" and gives her address and full phone number.

The Mexican Fish Hound

At the Baja camp where we have a little cabin, we dubbed the camp dogs Mexican fish hounds because of a unique task they had been trained to do. There are wide tide flats which are barely covered by water at low tide. Some rather good fish often get caught in small puddles here and the dogs had been trained to go out, find them, step on them and bark. The Mexicans then come along with a bucket and collect these fish. With the supply truck broken down and food supplies running out, I have seen the dogs literally catch enough fish to feed the Mexicans and sometimes also the restaurant customers! It is occasionally rather humiliating for American fishermen to return after a long day of unsuccessful fishing and be shown the large bucket of fish the camp dogs caught that day!

The "daddy" dog for a long time was a handsome friendly rascal named Chato (Spanish for "blunt nosed"). He was known and loved by many visitors to this area of Baja. He was a great swimmer and in the water as much as out. One evening when we were down there he got his hind leg hopelessly injured by a boat propeller. His Mexican owners were preparing to take him to the dump to shoot him when I suggested it would be worth the trial to see how he could get along on three legs. They readily agreed, so we fashioned a M*A*S*H style operating room on the beach and, with light from the headlights of a circle of dune buggies, amputated his leg. He was put to sleep with an old can of ether I had kept around for starting dune buggy engines. Its label said not to use it for anesthesia after it had been opened for 24 hours and it had been down there opened for months, but it still worked fine.

Chato was probably my most famous patient. People from everywhere kept calling my office to see how he was doing. The surgery was successful. On three legs he still swam vigorously and was observed to catch and kill jackrabbits and even a coyote! He sired a litter of pups which his Mexican owners felt was a magnificent display of *machismo*. I was given the pick of the litter and thus started the breed of Mexican fish hound.

Mexican fish hounds at work in Baja California. When food got scarce these dogs would literally catch enough fish to feed their owners.

CHAPTER 22

HEALTH AND FIRST AID

MEDICINES AND FIRST AID • WATER • MONTEZUMA'S REVENGE • FOREIGN FOOD • EAR PLUGS FLYING PREDATORS • CPR (YOU HAVE 180 SECONDS!) KEEP HYDRATED • HYPOTHERMIA • SCUBA AND FLYING

I was rather surprised to find that this chapter was hard for me to write. I had supposed that, as a physician, it would be easy to just dash off everything medically important to wilderness flying, but actually it is difficult to separate the really important items from the great many things in medicine that could have some application in this respect. Also, flying, exploring and writing about it is, for me, a hobby that is an "escape" from the discipline of medicine and there is a subconscious reluctance to bring my hard workaday world over into my world of escape. Not wanting to produce a thick medical book rather than an exploring book, I have followed the same philosophy as in the rest of the book. Thus, this is not a book on medicine in the wilderness, just as it is not a book about the actual technique of flying, but instead includes some of those items of medicine and first aid that are different from the usual because of being in the wilderness. In the case of flying, we have looked into situations that are different because of the wilderness's lack of many aviation aids we are used to in civilized country and similarly, in the case of medicine we will take up those aspects that are different and should be emphasized because of the lack of the usual medical aids. Again, as we did with flying, we will not go into many details about first aid or medical needs because there are already many good adequate books on this. Your survival manual should have an adequate chapter on first aid and for those who may want a concise digest of the latest in much of the medical field the Merck Manual is recommended. [23]

MEDICINES AND FIRST AID

Most pilots seem to be basically healthy; the periodic flight physicals weed out the seriously diseased ones. The typical pilot tends to be the gung-ho healthy type; flying doesn't usually appeal to the psychoneurotic stay-at-home-to-keep-the-feet-dry type of person. There are many exceptions to this, of course, but still, pilots do tend to be healthy goers and doers who are not overly concerned about their health. At the same time, ordinary flying today can be done, thanks to the comfort and ease of the modern lightplane and the comfortable pilot facilities at most general

aviation fields, without any excessive requirement for strength and stamina; and today, health in pilots is usually concerned only with any physical impairment that might interfere with adequate concentration and coordination for safe flying. The old days are gone when the pilot had to be strong enough to prop his plane, tail it around in a muddy field, agile enough to climb into the plane by stepping on delicate parts and avoiding guy wires, and have sinuses strong enough to withstand the rigors of the open cockpit.

But when you leave the relatively pampered situation of the usual general aviation facility and contemplate going out for a stay in the wilderness with your plane, other aspects of health become important. It may mean more physical exertion than you are used to to move the plane around, carry equipment or go hiking in the mountains. You become more reliant on yourself and your physical abilities. If you become ill or injured, medical care may be a long way off. If you have a tendency to such things as colds, sinusitis or serious allergies, you should consider that you could be stuck out for several days in bad weather. With proper clothing and equipment, you should be able to keep warm and dry, but there is still an exposure that is different from what you are used to. The wilderness is usually without medical help, and with a plane, getting to it may be delayed because of darkness or bad weather. So proper emergency medical supplies are much more important than they are for a trip where drug stores and doctors can be reached. What should be taken will vary depending on the potential needs of those involved and with the ability of someone along to diagnose medical problems and prescribe the proper remedies.

The usual first aid supplies should be supplemented with any medicines routinely taken and with what are likely to be needed. If you are taking passengers, be sure and stress to them the need for taking along medications that they will need. It is amazing how hard it is to make some people realize that there are no stores in the wilderness! If you get out there and suddenly someone needs something that was not brought along, it means that their discomfort may put a damper on the fun of the trip, or possibly you will have to fly to a place where supplies are available. The flight and getting into town and back will often shoot most of a day. If passengers don't have a prescription along, it may be very hard to get some medicines, and if you are in a different country, medicines may be known only by unfamiliar names.

The appendix contains a suggested list of items for a first aid-medical kit, but it is only a bare outline that needs to be augmented by your own individual requirements.[24] These items should be put in a sturdy container and kept in a specific place in the plane so that they can be quickly reached if needed. Have the first aid or survival manual there too.

Gamma (Hyperimmune) Globulin

If you will be going into underdeveloped areas where you will be in close contact with local people, there is always the chance of contracting infectious hepatitis. This is a liver infection by a virus that can cause severe long term disability and is occasionally fatal. While it occurs everywhere, it is concentrated in certain areas where sanitary conditions are poor. It can usually be prevented by a shot of "gamma globulin" which is a concentration of the elements in human blood plasma that contains the "antibodies" that fight off disease. Plasma from many people is combined so that there are disease fighting elements which were developed in the blood stream of many different people with many different diseases. It has also been used on statesmen and others whose health is important to help prevent colds and other infections. Its value for this is somewhat controversial but it is nearly always harmless to use. Whenever we are going into very undeveloped areas of Mexico for medical work we always take a shot beforehand and since I have been doing this I feel sure that I have had fewer colds and other miscellaneous infections. So, if you think you will be exposed, ask your doctor about the advisability of such a shot for you.

Water

Carrying drinking water can be a problem if you are near gross weight already. Even if you aren't, every pound of weight that can be saved is a safety factor, especially if you are flying in the mountains or in other areas with short poorly surfaced strips. Still, you must carry an adequate emergency supply when flying into wilderness areas. If it's to the American Southwest, Mexico or other desert areas, more will be needed than if it's to Canada, Alaska or other wet areas. Be sure water is carried in containers strong enough to survive a bad landing. The usual plastic gallon jugs of drinking water sold in grocery stores are made of very thin plastic that easily ruptures. Gallon bottles of liquid laundry bleach are much stronger and if saved and washed out make excellent water containers. If you know anyone who works in a hospital, perhaps you can get some empty rectangular plastic bottles that sterilized water sometimes comes in. They are slightly heavier but are much stronger. Their tops are very leakproof and the square size makes them easy to pack.

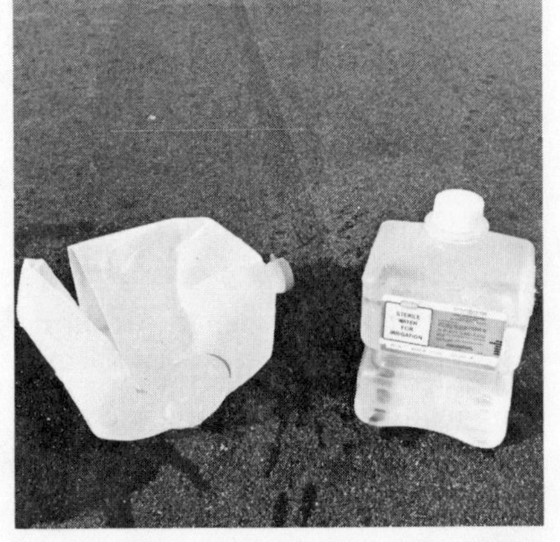

In an experiment by the author these plastic containers were dropped on cement from a height that would perhaps create the force of a "walk-away landing." The usual plastic water container ruptured while the square hospital container stayed intact.

"MONTEZUMA'S REVENGE" (DELHI BELLY)

In many areas, it is wise to disinfect any local water that you may have to drink. (Several simple ways to do this are listed in the appendix.) [10] Infections from contaminated water supplies usually cause a relatively temporary illness. It is often treated lightly and given various semi-humorous names. Unfortunately, it sometimes is not humorous. Some of these infecting organisms can cause liver abscesses and other serious conditions which can result in long term disability and even death. And even if you don't get one of the bad bugs, the temporary ones can make you lose a day or more of your vacation.

So, use local water and ice with care. Toothbrushing is an often overlooked source of local water infection. "Bottled water" in many places is just a bottle that is filled in the back from the local water. If you are stuck, soda or any soft drink that bubbles when it is opened is safe. But watch the bottle tops. Often they are kept submerged in local ice water and this can collect around the edges of the cap. Less than a drop of the contaminated stuff is all it takes! Freshly opened beer or wine is safe too, and some travelers feel compelled to save themselves this way.

Treatment

Treatment has two parts: Slowing the excessive intestinal activity and killing the bug. Diarrhea and cramps can be helped by "Lo-motil" tablets (two-four times a day, maximum of twelve in 48 hours) or just plain atropine (belladonna) tablets (one tablet four times a day). Stop either of these if your mouth gets too dry. "Kao-pectate" is another time-tested remedy that helps. Normally it is a liquid, but for flying a more concentrated solution ("Kao-con") or, preferably, tablets are available. These work just as well and save a lot of weight. In severe cases, nothing gives the comfort of good old fashioned paregoric (denatured tincture of opium); it slows the intestine and also is one of the original tranquilizers. (Don't fly after taking it, and be sure any bottle of it doesn't mention opium if you will be crossing any international borders.) Once you begin to get relief from any of these, taper off to just enough to keep you comfortable.

To "kill" the germs, antibiotics or sulfa drugs may help. It depends on what medications the germs are sensitive to, and this is impossible to determine in wilderness conditions, or indeed, in even the best of medical laboratories. The cause of travelers' diarrhea is a puzzle that has baffled many medical investigators. [25] The war in Vietnam stimulated much intense and high-powered research without results. In many cases no specific germ can be found and often antibiotics don't work. Not all my colleagues agree, but, just to be sure, I personally usually use a sulfa drug and a so-called "wide spectrum" (Shotgun type) antibiotic that is effective for a great many different bugs. Whatever you take along should be reviewed with your doctor because they can have side effects on certain medical conditions or can mix poorly with medicines you may already be taking. Be sure that your doctor understands what you want them for; many city physicians have had little experience with the treatment of the contaminating organisms found in primitive places.[26] If you may have to fly while taking the medications, he should know this too, as some of them can affect your reaction times, etc. Even if you do not contract any actual infection from local water, your system can still become upset by the change in water and especially by repeated changes during a rapidly moving trip. It's not that there is anything wrong with each water supply taken individually, but the change in mineral content from place to place can do it. You don't have your usual zip and you don't enjoy your trip as much as you could. If plans call for hiking or other more strenuous activities, the effect can be increased. It's a variable matter from person to person, and only you know how delicate your tummy is about such things and how important an unchanged water supply is to you. If you have the weight carrying capacity, it may be a good idea to carry enough water for as long a period as you can, to minimize the water changes. Even the disinfecting agents can be somewhat upsetting to many people.

FOREIGN FOODS

Much that has just been said about water can be applied to foods in foreign countries. Like water, foods can affect you by being "contaminated" and causing infection or by just being different from the usual diet. Foods to be careful of, are those whose outside surface has been handled in its picking and distribution. An apple or a tomato, for example, should have its skin well cleaned before eating, while a banana or pineapple, if carefully skinned, would be safe. Some investigations have suggested that it is really a matter of climate or some other as yet unknown factor because travel "horizontally," i.e. east–west, usually gives no problems, but travel "vertically," i.e. north–south, anywhere on the globe can be associated with difficulties. There is a great deal of psychology involved here too: Many people who would never think of eating anything in a foreign country, eat vegetables at home all winter long, not knowing that many of them are imported from Mexico.

Long ago when I was a medical student working in the Emergency Room of the Massachusetts General Hospital, I learned to have great respect for the power of food poisoning because we would be periodically swamped with a large number of very acutely ill patients who had eaten that day at one of Boston's otherwise very superb seafood restaurants. This fearfulness naturally carried over to the occasionally contaminated Mexican food and especially Mexican seafood. On one Mexican trip a companion took me to a very clean and spotless restaurant in a little town, but I only watched while he consumed the tasty Mexican cooking. Later I ate a sandwich from home. A little further along we stopped at a primitive little town in search of live lobsters to take back home. We found a picturesque character who said he could furnish us with some but wanted us to sample his product first. He took us into his humble kitchen and on the back of a dirty stove was a large kettle containing a number of cooked lobsters soaking in a scum-covered solution of tepid water. He put his dirty hand into this and pulled out a cooked lobster which he handed to me as he rummaged through a filthy drawer to find me a somewhat-used fork which he cleaned off on his dirty pants. My companion, who had been chiding me for my finickyness at the clean restaurant, was very amused and whispered to me, "He will be very insulted if you don't eat it!" Well, I did; it was good, and we bought some excellent live ones and nothing untoward happened.

"Food Poisoning"

A frequently overlooked hazard for a pilot, not only in foreign countries but also in flying anywhere, is the ordinary "food poisoning" which occurs with poorly refrigerated products, especially those containing cream or mayonnaise. Potato or chicken salad, for example, left outside on a hot day (and perhaps walked upon by a few flies), is notorious for causing wholesale epidemics at picnics and other outings. These attacks are rarely long lasting or serious, but can be very hazardous to a pilot because of the sudden and severe onset, frequently to the point of almost total incapacity to fly. Pilots who have had such an illness, or know of someone who has, become very careful of what they eat for some time before a flight. Picnic type food carried in the plane should be well refrigerated or else composed of non-infectable items. These diseases are usually "self-limited" and not serious. Their danger is in the sudden onset of almost complete paralysis in perhaps the only one who can fly the plane.

EARPLUGS

One item that can make a big difference, if not to your actual health, at least to your comfort, is ear plugs. On a longer flight one of the greatest causes of fatigue is the noise. You may have discovered this already if you use padded type ear-covering headphones. Yet, in wilderness flying the radio is used so infrequently that there is often no need for wearing headphones. Earplugs will give the same, or even better, noise-cancelling effect and can be used by passengers too. Many types are available; be sure to get a type designed to minimize *all* noise, not just severe sudden noises such as shotgun blasts. My preference is for the small rolls of foam material that are compressed to insert in the ears and slowly expand to really close out the noise. It feels a little funny at first but you soon get used to them. If you haven't tried ear plugs – you should! They are inexpensive and each person should have his own pair to avoid transmission of ear canal infections. Many people have little fungus organisms growing in the ear canals quietly and without symptoms. If they are transplanted to another person, however, they may become more active and cause discomfort. If you get used to earplugs for flying, you may find that you will use them for other things too. They are great for shutting out the kids' stereo in another room and the other night I found my wife wearing her's – she thinks I snore!

FLYING PREDATORS

Insects can be related to health on wilderness trips, but with some rare exceptions those encountered on this continent will not be carriers of disease.

Flies

Unlike some other bugs, flies do not get sick and transmit their diseases, but they do carry germs mechanically. In areas of poor sanitation they can bring dysentery and possibly other illnesses if they walk around on food or dishes after flying in from some nearby source of infection. Fly control is a longer range program than you will probably have time for. The several forms of DVDP, (used in fly strips, flea collars and as a spray) do eventually kill flies (and other bugs) in an enclosed area, (although resistant strains seem to be starting to appear), but this chemical is a slow-actor and is useless outdoors. Many rural people still use the old sticky fly paper rolls, but these too are slow and are very unwieldy for camping situations. Household cans of aerosol insecticide will work quickly to kill existing flies and, to a lesser extent, to repel newly arriving ones. These must be used with real caution around food areas – the ingredients can be more hazardous to your health than the bugs. The danger from flies is best combatted by keeping them off food areas by various coverings or screening.

Ticks

Ticks don't fly, but in some areas of the west they can carry serious infections, and it is a good plan to inspect yourself (and your animals) well after being in the brush. Look for the little brownish crab-shaped creatures. If they are pulled off, the head often remains buried in the

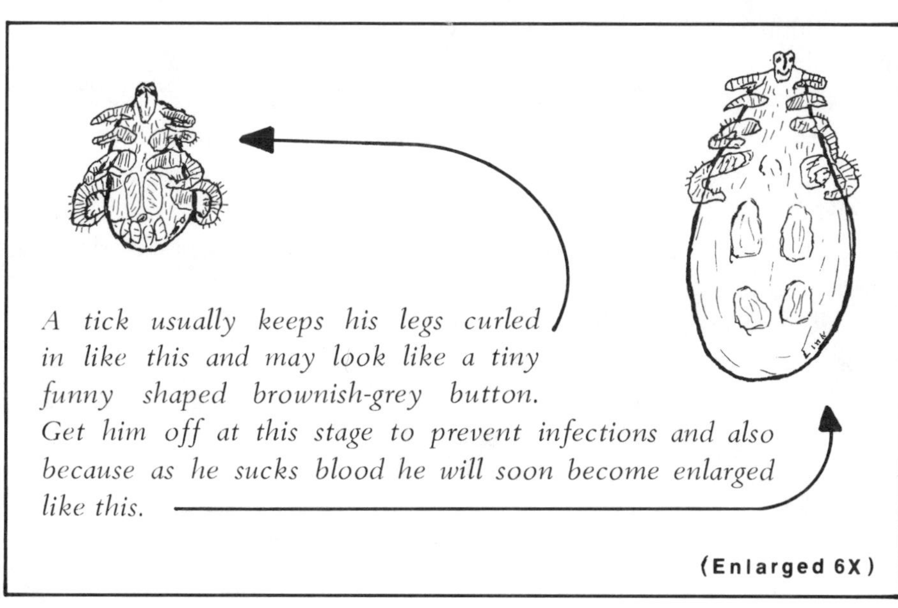

skin. A better way is to soak a cotton swab in insecticide solution and soak the tick with this. Wait about a half hour and then the entire tick will come off easily.

Mosquitos

But the real problem insect, and it's more of a nuisance than a danger, is the ubiquitous mosquito. In summertime, in any areas that have a significant amount of rain, they are usually a problem. The number and size of Alaskan mosquitoes is legendary, but their bites are not as toxic as some of the smaller mosquitoes elsewhere. Their bites itch for a relatively short time as compared to, say, their smaller relatives in New England. In many parts of the country, mosquitoes are a part of life, and most people become more or less expert in avoiding them. Those who live in California and other fortunate places are often very naive about mosquitoes and how to live with them. In mosquito country it usually pays to keep as much of the body as possible covered; long sleeve shirts and long pants and, in extreme cases, hats and mosquito netting over the face. There are unsubstantiated tales of severe injury to exposed parts of the body from relatively short exposure to large swarms of mosquitoes. They like to hang around brush and deciduous trees, so staying in open fields or other relatively bare areas may help. Winds will keep them away and so will campfire smoke, although sometimes being in the smoke is more unpleasant than the mosquitoes. Modern science has come to our aid with increasingly effective insect repellants. The old time "oil of citronella" of pre-war days was a help and at least had a pleasant smell. However, wartime research and subsequent improvements have devised more effective chemicals that are usually harmless to the skin (although one *can* be allergic to almost anything) and make insects loose their interest in you. There are several brands and each has its devoted enthusiasts, but the ingredients listed on the containers all show about the same things. Creams, aerosol sprays or concentrated drops – use whatever seems most satisfactory.

A good routine is to always keep a small dropper bottle of your preferred repellant in the side pocket of your tent. Also, be sure to *always* keep one in a side pocket of the plane. Sooner or later, you will be on a side trip or somehow away from the gear containing the insect repellant, and the plane's supply can make all the difference in enjoying the side trip or not. Be sure that the tent has a fine mesh mosquito-proof net screening all openings. Even so, some will get in when the flaps are opened, even if ever so briefly, so you may

need to use repellant on yourself inside the tent anyway. Other possibilities are a can of household aerosol type spray – a short blast into the closed up tent a little while before you are going to use it. The use of mosquito coils, a circular punk type arrangement, is effective but somewhat awkward to use in a tent. The advantage of killing the mosquitoes with these rather than using repellant is that with the repellant, mosquitos may not light on and bite you, but they are still aggravating as they fly noisily around your ears trying to figure out why you smell so bad to them.

CPR. (YOU HAVE 180 SECONDS!)

Every discussion of first aid should include a strong recommendation to become trained in cardiopulmonary resusitation, (CPR). This is very advisable for everyone everywhere, but in an isolated area and especially in a camping situation, where there are likely to be lakes and rivers, it is a must. This rather simple procedure is an updating of the artificial respiration technique. Artificial respiration alone is very valuable in cases of drowning, etc., where breathing has stopped but only when the heart is still beating. CPR extends the effectiveness of the resusitation procedure into the several minutes when the heart has stopped but can sometimes be restarted. It is quite simple and easy to do, but it must be done with some exactness. Therefore, it cannot be learned properly from a book, but needs a short period of personal demonstration and instruction. Doctors, lifeguards and other medical personnel are usually required to take a refresher course every year, and this is a good idea for everyone. It will not always work, but if you have been able to do it properly, you will have the satisfaction of knowing that you did everything that could have been done. If you do not know where to get instruction, your local physician or the emergency department of your hospital would be logical places to inquire.

KEEP HYDRATED

Recent medical research has shown quite conclusively that at altitudes over 5,000 feet the human body loses an excessive amount of water through the breath. This applies to pilots of lightplanes above this altitude, as well as to all jetliner passengers. If the flight extends for several hours, the body can become noticeably dehydrated. Preventing this by drinking a lot of extra water or other fluids during the flight will cut down on fatigue, give better coordination for flying and make one feel better the next day. (While thinking about hydration, if ever stranded in the snow try not to eat it. If at all possible melt it into water and drink it. The body uses up an excessive amount of energy in melting eaten snow. If clean ice is available it will melt using less fuel than an equivalent amount of snow.)

HYPOTHERMIA

Any reading or discussion about camping today soon turns up the word "hypothermia." This is a relatively new and often loosely used word which literally means the cooling of the *entire* body, i.e. the "central core," as it is called, and not just the skin layers. It started out as a purely medical term describing a new technique in which the entire body is very carefully cooled several degrees as an adjunct to anesthesia during certain very complicated surgical procedures. The use of the word has been extended to describe the cooling of the entire body from other causes, primarily exposure to prolonged cold weather outdoors while inadequately protected. Here, it is a very dangerous condition, because the body makes extreme efforts to maintain its normal temperature and lets it become lowered only as a last resort when its defenses are becoming overcome. It is literally the first stage of freezing to death. Treatment must be prompt and vigorous and is most successful when done at the first onset. However, at this stage the victim is often not aware of his condition, so if you are in a situation where hypothermia is possible, watch your companions carefully because it often starts only as vaguely strange and inappropriate speech and behaviour.

Since the word has become popular, it is often used quite freely to describe conditions that are not technically as serious as actual body temperature lowering. It is like the word "shock" which has specific (and serious) medical meanings but is used popularly for all sorts of relatively minor conditions. Similarly, anyone who has had the skin only seriously chilled, with perhaps some frostbite, is often described as suffering from "hypothermia." *True* hypothermia is relatively rare, but it is a serious medical emergency, a rapidly progressive condition that is usually fatal.

The treatment is, naturally, to warm the body back up. Warming the body's outside surface may increase the hypothermic condition on the inside where it is most dangerous, so, unless the victim is unconscious, it is best to start treatment by giving hot coffee, tea or soup.[27] At the same time start warming the skin. One of the fastest ways to do this is to remove all clothing, dry the skin if necessary and put the person inside a sleeping bag next to another warm bare body, or two if there is room. Skin to skin contact gives the quickest heat transfer. This is a lifesaving procedure and is no time for modesty or giggles. Other warming methods can, of course, be used, and a fire, or a warm room with plenty of blankets, jackets or other insulation should be used after the warm up has started. Like so many things, avoidance is by far the best treatment. With proper equipment hypothermia should not happen.

SCUBA AND FLYING

The lightplane is very popular with scuba divers for getting to otherwise almost inaccessible spots, but don't forget that flying altitudes soon after diving can increase decompression problems.[28]

Bear in mind that good scuba (or snorkeling) requires a combination of two things. (1) Clear water and (2) something in the water to see. If you are a novice at these things, don't fly off to somewhere only because of reports of very clear water. (A bare sandy bottom seen through 30 feet of sparkling clear water looks no different than any sandy bottom through two feet of ordinary water.) There must also be interesting marine life to look at. Usually this requires the presence of a coral reef nearby or else rocks or marine vegetation for fish and other things to live around.

PART IV
HOW TO PLAN AND WHERE TO GO

CHAPTER 23

PLANNING A TRIP

HOW LONG AND HOW FAR • KEEP IT LOOSE • ALLOWING FOR WEATHER • REVERSED PLANNING

Okay – you're sold on the adventure and excitement of a flying vacation into the wilderness. What do you do next? Why not just put together all the suggestions in this book and start out? Unfortunately, it's not quite that simple. Almost, but not quite.

HOW LONG AND HOW FAR?

First, of course, you have to decide on the amount of time that you can or want to take for the trip, and from this you can calculate a ball park figure of how far you can go. You may already know how much flying you can comfortably do in a single day, but unless you have already taken trips of several consecutive days in a plane, you may not know how you will tolerate this. It's a very individual thing. Some pilots seem to be indefatigable and can remain glued in their seats hour after hour, day after day and thrive on it. Others are less rugged and find that six or eight hours of straight flying leaves them a little pooped the next day. Other things will make a difference too. Emotions are tiring. If there is anxiety or fear, it can wear you down much faster than straight flying. If you get lost and nervous about it; if weather is worrysome; if a head wind is eating up gas faster than expected – things of this sort can use up great amounts of emotional energy and tire you faster than usual. On an extended trip, it is possible to "overdraw" just a little each day on your energy supply without noticing it. After several days, these little daily deficits can suddenly catch up with you as a surge of overwhelming fatigue. Often this fatigue masquerades as a somewhat vague illness. This is common to all types of travel; but is much more likely to occur when you have to navigate and fly the plane and especially when you have the added responsibilities of providing food and supervising camping.

My own answer for this has been that when I am on an extended trip and I find that I am at a spot that is interesting and enjoyable and the weather is good, I stay over an extra day. No flying that day. The bird and I both rest. Try it. Put on a dark sleepshade and "sleep in" as late as you want. Or, if your system prefers, you may get more relaxation out of a day by getting up relatively early and wandering around the area, exploring, fishing, gold panning, hiking or doing whatever the spot provides. Then back to camp for a drink, a light lunch and a long afternoon siesta. This recharges my battery more than anything I can do.

KEEP IT LOOSE

One of the most restful aspects of this type of trip can be the lack of a need to plan very exactly. When you are camping, one of the delights can be a complete indifference to any schedule. Especially if your regular work is rigid and has a demanding schedule, you may find the looseness of an unscheduled trip extra appealing. Or, perhaps scheduling has become so inherent in your make up that you are uneasy without it and are happier to have a fixed timetable for your trips. No problem. It's your trip. Enjoy it in the way that is happiest for you.

But be sure to consider carefully that one of the most enjoyable features of this type of trip can be the ability to change and adapt the schedule to fit your own desires as you go along. You probably don't have much chance to do this in your workaday life; but here is where you can, so take advantage of it. On any trip it is very easy and natural to get into a mental set where you feel that you "have" to be a certain places by certain times. Everyone has had the experience of stopping at some very appealing place and having to leave sooner than desired because you "have" to get to Horseshoe Swamp that night because that is supposed to be a super place to relax. Well, you know it! Horseshoe Swamp turns out to be a hole and you wish that you had stayed at that nicer place that you hurried through. Somehow the grass always promises to be greener farther along. There is a psychological fix that the farther you go the better things will be. Well, not so many times! You may be at the peak now, so stop and enjoy it. No one can travel for anyone else, so don't rely too much upon someone's report of an excellent place ahead. If you find your own excellent place stop and enjoy!

Pilots vary in their need for schedule. Some have a compulsion to fly with their computers in their laps; they constantly know the wind at their altitude down to the last knot; their ground speed is reviewed at frequent check points; they are human DME's; constantly they fiddle with fine adjustments of power while worrying about cross wind components. Others jump into their planes as if they were automobiles and simply fly to their destination, watching their navigation only enough to be sure they are roughly on course. If they know that there is enough gas and that the plane is functioning adequately, that is all they want to bother with. Most of us fall in between these extremes and as long as fuel supply, engine function and navigation are adequately monitored, it probably doesn't matter too much either way. This is contrary to what is taught in flying schools and advised by the FAA (which provides a form for exact check point times and distances on the reverse of every flight plan form). However, this much work and detail is not necessary for a safe flight. You must be very sure that you are always within a safe distance of needed fuel; you need to keep track of your general location to give position reports as well as to keep from getting lost; and you have to watch upcoming terrain elevations. But you can relax and have a perfectly safe and a very enjoyable trip watching the scenery with the stopwatch and calculator left at home.

So plan, but only to the extent that you desire. Many of my best trips have had only very vague planning. I would know of a place I hoped to reach; but I was more interested in seeing what I could find along the way. Each night, I literally didn't know which way I would go the next day. Often it depended on the morning weather or upon some whim of the moment. When you are planning things ahead of time at home, your decisions are made on rather abstract impressions compared to what you actually find when you are en route. Along the way you will often pick up local travel maps and folders that will show things that you hadn't heard about before. People you meet on a trip can fill you with enthusiasm for a detour to see something interesting. So the best planning may be practically none, except for a general study of the area you hope to cover.

This study is usually best done with ordinary travel materials, rather than aviation sources. General travel guides, AAA guidebooks, the National Geographic Magazine and the like will help you in picking out an area to visit, will provide information about various things to see there and in general will give a better overview of the area to make the trip more interesting.

It is hard to tell ahead of time just how much you will have the time and energy to see. I try to calculate very conservatively how much I think I will be able to see; but I usually end up by being able to cover somewhat less than half of what I usually estimate. It's good to have the planning over extended in this way, because it is better psychology to have to leave things unseen while you are still interested. You can always come back another time. The other alternative, of being underplanned, could leave you with nothing of interest to see or do.

ALLOWING FOR WEATHER

How much time to allow for possible bad weather is a problem in wilderness flying, just as it is in all private flying. This is where some over-extended planning can help. If the weather delays you, the extra itinerary can be omitted. If you are lucky and have minimal weather delays, you can go ahead and visit these extra places.

The actual amount of time that should be allowed for weather is dependent on many variable factors and must be left to your judgement in each individual case. The thing to remember about wilderness flying is that if you get stuck by weather, there may not be commercial transportation to bail you out!

REVERSED PLANNING

As flying is usually taught and practiced, a flight plan starts with selection of airports and a route is then planned between them using aviation charts. For our type of trip, planning should be somewhat reversed from this. It is better to plan with general travel maps first and fill in the aviation details later.

For the preliminary planning of a trip, ordinary geographical atlases or regular road maps usually show points of interest better than aviation maps. Start with a large scale road map. (There are some that cover half the country, or have most of western Canada plus Alaska on one side.) Then zero in on specific areas with ordinary road maps. After you decide where you want to go and what you want to see, the flying details can be filled in later from aviation charts.

AAA maps and travel guides are an excellent source of interesting things to see and do in any area. Points of interest are well described on one section, and the listed accommodations in another section can often indicate even more about the general character of a resort area, especially in relation to the amount of wilderness present. Often the dates of the busy season are shown, usually coinciding with the time of best weather. And, if you are a true wilderness flier and want to get away from crowds, this tells you when not to go there. Often just off season the crowds are gone but the weather is still good. But in many ways the AAA listings are the very antithesis of wilderness concepts. They list only what they consider are the top seventh of all accommodations, and these are usually the most luxurious and least isolated ones. Adventures in the wilderness usually just begin where they leave off. If an area or resort is listed there, it is almost automatically not a wilderness spot as I have been defining it.

Trip Planning

So, decide where you want to go and put one of the large scale road maps on a table. With a yardstick, draw a pencil line between the final destination and home. Then look at things that you might want to see on both sides of this line, remembering that with a plane you can disregard where the roads go. Take the smaller road maps of each state or area and look for more detailed information about possible good side spots to see. Use the AAA guide books and other sources and read about these areas. Now, make a new pencil line connecting the chosen places that you would like to see. Then get out the flying charts. *(Sectionals, WACs or state aviation maps are better than IFR charts for this.)* Find airports near to where your points of interest are. Look them up in the AOPA Airports, U.S.A. for information and to see what ground transportation is available if the airports are not right at the point of interest. Finally, refine your actual flying route by studying the air maps for geographic and geologic features that should govern your specific flight route to minimize mountain flying, etc.

CHAPTER 24

HOW TO FIND A GOOD WILDERNESS STRIP

MAP AND BOOK WORK • AIR OBSERVATION
GROUND INSPECTION

Extending the planning to find wilderness strips and good camping strips involves three phases: *Map and Book Work, Air Observation,* and *Ground Inspection.*

MAP AND BOOK WORK

On the sectional or WAC chart scan the area you are interested in with special attention to the following:

1. Forget the solid blue and magenta colored airports with control towers listed. A lot of these will permit (or tolerate) camping; but if there is enough traffic to warrant a control tower, it is too much for peaceful camping.

2. The same *may* apply to any airport with a cross extending beyond the circle, indicating fuel and other facilities. This almost always indicates an operator in attendance and probably enough planes and people to make camping a rather public performance. There are many exceptions to this and often there are areas off side runways or unused taxiways that are satisfactory. The operator of a small airport is often a little lonely, especially on weekdays and is usually interesting to talk to. Often you will end up getting invited home for dinner.

3. The most satisfactory wilderness and camping airstrips are those without any facilities, usually those indicated by open circles (non-paved) but sometimes by solid circles (paved), in each case without the cross extending beyond the circle.

4. Study the *surroundings* of any strips that you have selected. How close is a road? Practically all strips have some sort of road going to them, if only for the construction and maintenance of the strip. Dozers can be airlifted in, but these cases are extremely rare. But how close are roads other than those just to the strip. The sectional should show something of the quality of the road, either a rough trail (dotted), minor road (thin line) or regular highway (thicker line). The amount of traffic that could be expected on a nearby road can be estimated from this as well as the relationship of the road to adjacent towns. Is it a main route or are there

better or more direct routes between towns? The elevation of the road as compared to the strip may be significant. (A road in a valley beside the strip may not be seen or heard from the strip.) Also look for other topographical features such as lakes and rivers and how close they are to the strip. Surrounding mountain scenery can be estimated by the contour lines.

5. The *elevation* may be significant. Higher altitudes will generally be cooler, which may be good or bad depending on the time of year and your personal preferences. In winter or spring the snow level and often the condition of the strip surface is often related to the elevation.

6. Other miscellaneous characteristics of the area may be suggested by the map, possible windy conditions from adjacent mountains or mountain passes, dryness indicated by nearby dry lakes, etc.

Strips Not On The Map

There are many strips that are not shown on the maps, many of which are in isolated areas and have good camping potential. Often these are farm strips, and if the farmer is a pilot he can probably be easily approached and may be willing to let you camp adjacent to his strip. If it's on the "back forty" this is fine, but if it is near the farmhouse, camping might be a little awkward. Even if you don't stay, many of these flying farmers are always happy to have someone drop in to hangar fly and many of them make very interesting acquaintances.

Additional non-mapped strips have been built in connection with other enterprises, such as mining, logging, dam construction, firefighting, or commercial fishing. Often they are quite long and wide because they were designed for cargo aircraft to bring in supplies or take out cargo. Frequently, the purpose for which they were built has ended, and they remain there essentially unused. The problem is that if they are no longer maintained, how good is the surface now? Especially after rainy periods or spring thaws, rodent activity and other deteriorating influences, the surface can be very treacherous. (See pages 24 to 26 for suggestions about evaluating such airstrip surfaces.)

AOPA Airports USA

Planning done at home prior to the trip should produce a number of promising strips which can then be further evaluated with the AOPA Airports USA. This computerized listing of airports is published yearly by the AOPA and is an indispensable aid in evaluating airstrips in the

United States and Alaska.[29] It gives information that can be obtained in no other practical way. Occasionally it takes a little ingenuity and imagination to find the listing because it is not always listed under the name that the sectional gives. It may be listed under a nearby city or town (not necessarily the nearest town) or by the county. But nearly all of them are there if you can find them. It will indicate the runway direction, length and surface as well as fuel availability, repair facilities and other services.

During a trip, this same planning with maps and books can be done in the air if it is getting time to stop somewhere.

AIR OBSERVATION

The next step is to look over a strip from the air. This will repeat many of the things that were considered when checking the map, and most of the same features are reevaluated. The strip's condition is correlated with sectional, and sometimes significant differences are found.

Things may have changed since the sectional was last updated, or there may be an error in the mapping. Changes in bodies of water are frequent because of a recently built dam. (Many beautiful stream valleys are now under the surface of a reservoir.) When compared to the mental reconstructions made after looking at the sectional, the actual appearance and features of nearby mountains may be quite different. The size and nearness of towns, buildings or roads may also differ from the map.

As you fly over the strip take all these factors into account. Look especially for indications of activity adjacent to the strip, especially nearby residences or other indications of human presence. These do not necessarily rule out the possibility of a good camping site, but from the air you can get a good idea of whether the whole strip is isolated or has isolated areas near it, or if you would have to camp in someone's backyard. Such a flyover will rule out many strips. A flyover cannot always be counted on for a final decision, and I have been fooled from the air both ways. Some strips that look excellent from the air are not so on the ground, and some strips that seem very questionable from the air make excellent camping sites.

GROUND INSPECTION

On the ground is, of course, where the final decision will be made as to whether or not you want to spend some time there. (Before landing for a ground inspection, the strip should be checked out for safety for landing and takeoff as described on pages 24 through 28.) You can start the inspection as soon as the landing roll is under control. As you roll, look at the surroundings of the strip. The far end may be the best place for camping, so look at that while you are down there. Are there level, non-rocky places along the sides of the strip far enough from the runway to be safe to park the plane? Can you safely taxi to these or is there a berm, rough area or rocks in the way? As you taxi back up the strip, continue this inspection. Are there trees or bushes you could camp behind to give you privacy from activities at the other end of the field? By the time you are back to the touchdown end of the strip, you should have a pretty good idea of the general suitability of the strip. Often you will have seen enough to definitely rule it out, and you can continue taxiing for take off to go and look for another possibility. If it still looks promising, park the plane near the most desirable area and get out for further inspection on foot. Be sure the area is safe from rocks, holes and hidden obstacles before you taxi. You may need to stop and check it out on foot first. Walking around will give you much more information. The surface may be too dusty for comfortable camping. There may be other undesirable features such as odors from trash, mosquitoes or flies; actual or potential noises from nearby generators, powerplants or roads that were not apparent from the air. Sometimes the place may not have any really specific bad points but may just not be a place where you have any desire to stay. The proper place will have a positive feel that will definitely appeal to you – the smell of pine trees, mountain scenery, the call of a nearby lake or river, or places that are inviting for exploration or hiking. If you have found a good place, it should be obvious to you at this point. If it does not really appeal, take off and try another. Usually, if a place comes this close to pleasing you, there will be others within relatively short flying distances that will be satisfactory.

Of course, your standard of approval will vary. If you are planning to stay for several days, you will want to be sure that the place is very good. But if it is late in the day, or if you are tired, a place can often make a very satisfactory overnight stop even if you would not want to spend much time there. Go to bed early and get up early and go and find a better place.

This process of trying to locate a spot first on paper, then by air and finally on the ground, is pretty much a trial and error method. Do not be discouraged if it takes a lot of trials at first. When you do find a good spot it will be worth the search. All of us who have done this have been discouraged by the number of duds that are found, especially at first, but, with experience, you will soon start to have a collection of favorite places of your own, and as you collect these you will be learning how to spot satisfactory places better. With more experience you can more quickly rule out unsatisfactory places by the map and from the air, and have more time to concentrate on more likely ones on the ground. But always keep an open mind. No matter how much experience you have, you will still find that from the air an occasional very dull looking strip will be great on the ground. There is much on the ground that just cannot be determined from the air, so keep a curious attitude and don't always rule out the questionable possiblity from air inspection alone.

Get all the suggestions and advice about a strip that you can, but do not be disappointed if you do not find it as described. One person just cannot travel for another, and no matter how well you may feel that you know someone, his recommendations for travel (as with his recommendations for movies or women) often will not coincide with your tastes. A fascinating archeological site to one is a garbage dump to another, so look over recommended spots but with reservations. The reverse is equally true and a place turned down cold by some friend may be an excellent one for you.

CHAPTER 25

A SPECIFIC EXAMPLE

In order to demonstrate in a specific way the off-the-pavement planning procedure described, here is a step by step outline of an actual episode during a trip made in late May 1976.

We are heading south in Washington and Oregon on a beautiful VFR afternoon, flying very close to the spectacular Cascade Mountains watching the spectacular isolated peaks go by about every fifty miles. As soon as one fades behind us, another appears on the horizon ahead of us. The air is so calm that we fly close enough to each peak to wave at the skiers on them. For us, it is a lazy man's mountain climb. The afternoon is wearing on; we are beginning to get tired, and it is time to start looking for a place to stay overnight. There are no clouds. We have been monitoring control towers as we pass within radio range, and no significant wind has been reported, although we would have known this from the very calm air as we passed close to the peaks. There is adequate camping gear and food aboard and we decide that it would be a fine night to camp at some quiet strip, and we start to look for one.

The *Klamath Falls Sectional* shows several nearby strips that might be good possibilities. We are tired and even though our equipment is adequate for it, we do not feel like staying in the snow that has been apparent in many spots higher than 3000 feet, so we rule out *Crescent Lake* at 4300 feet. Another candidate that looks promising on the sectional is *Santiam Junction State*. We fly over it and it looks isolated and quiet, but we can see a lot of heavy highway repair equipment parked on a road right next to the strip. They will probably have their heavy diesels going early the next morning. Also there are still blotches of snow in the surrounding country which indicates colder temperature than we want tonight. The sectional shows another possibility, *MacKenzie Bridge State*, so we fly over that. We look at it from 7500 feet because, if we don't land there, we will want at least that altitude to continue the flight across the mountains ahead. It

looks very promising, although from up here it surely looks small and down in a hole between all the mountains. We decide that it is worth a closer look. For the sake of our ears and engine, we descend relatively slowly, using this time and altitude to fly downstream and back for several miles to look over the surrounding countryside and see if there is anything that might be interesting to visit by land after we get there. We notice a small sleepy looking village, a rather large river and a highway. The highway and village seem to be far enough from the strip, and the river looks interesting. Looking in the AOPA Directory we are advised to land to the east and take off to the west. Is this because of adjacent hills, a sloping runway, or what? Closer inspection shows both; a slope up to the east and hills at the east end. We decide that the plane with its load could, if necessary, make a go around and have enough width to turn safely beyond the east end. If it were hotter weather or, if we were nearer gross weight, we would be much more cautious at this point. If there were any significant wind, the several narrow mountain passes coming together just east of the strip could increase the wind and make it very erratic.

All aspects of our aerial inspection have confirmed the sectional's indication that this should be a satisfactory strip. We land easily, although on roll out we realize that the strip is longer than it had seemed from the air and we didn't need to have come *quite* that close to the tops of the pine trees on final approach. Still we were perfectly safe all the way with very adequate safety margins. The turf field is rather bumpy, and we hold the stick back, to be as easy as possible on the nose wheel as we finish the roll and taxi to a parking place which is flat and well cleared.

As we get out of the plane, we are immediately entranced with the forested surroundings, the aroma of the pines and a distant wood fire. Further inspection confirms the suitability which was indicated from the air. The highway is not used much this time of year, and it is far enough away so any noise from it won't bother us and so our dog is not likely to wander over to it. The river is close by and beautiful. There is no one and no buildings to be seen anywhere. A camping table and fire ring right next to the parking area indicate that camping is okay. Pitching the tent and unrolling the mattresses and sleeping bags takes less than twelve minutes. (It's easier and faster to do this before it gets dark.)

Darkness comes late in this region at this time of year, so the remainder of the day is spent enjoying the peaceful river and surrounding countryside. The dogwood trees are in full bloom throughout the forest and are beautiful. A short ride by Mo-ped to the picturesque little village of MacKenzie Bridge shows us the local sights and gives an opportunity to buy a few fresh provisions to supplement our camping supplies.

The weather is so beautiful that afternoon that it never occurred to us to get the next morning's forecast when we closed our flight plan with Eugene Radio. (This we did while still at altitude over the strip to let them know where we were landing.) When we awake the next morning the Oregon weather has reversed itself and there are low clouds and fog halfway down the adjacent mountains. Last night we had set our altimeter to 1620 (the strip's published elevation) and it now reads 1635, which indicates no major barometric change and that the clouds are probably only a local condition. We doze on comfortably, peeping at the weather from time to time. The clouds are staying at the same level and there is no increase in

the fog. Down the valley to the West there is a light yellow tint to the horizon and the sectional shows a continual decrease in altitude all the way down the valley which constantly widens and contains a good sized highway. We decide that it would be safe to go and at least look a little way down the valley. Conditions continually improve as we go further down the valley and the weather is excellent before we reach Eugene.

This example shows a specific case in which we found a good overnight camping strip by first selecting possibilities on the map aided by the directory, then by checking them out from the air and finally, after a careful descent, by concluding the decision on the ground. The use of the altimeter to help judge questionable weather and the process of cautiously going up to look at marginal weather from the air were also illustrated.

USING THE ALTIMETER AS A BAROMETER

At isolated spots where no weather reports are available, the altimeter can give an additional weather clue by serving as an approximate barometer.

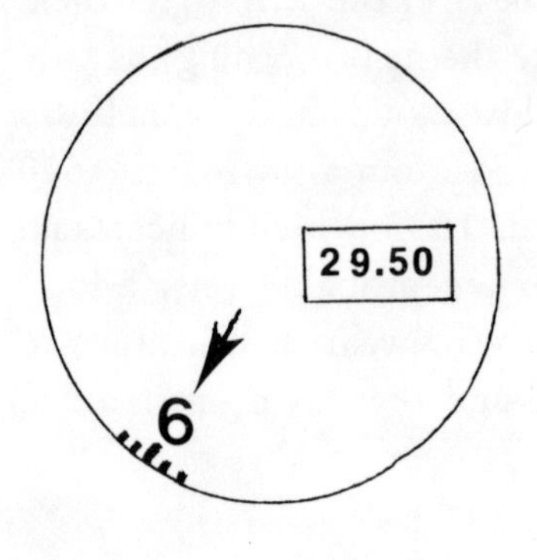

After landing, set altimeter to published elevation for strip, (or else write down indicated altitude.)

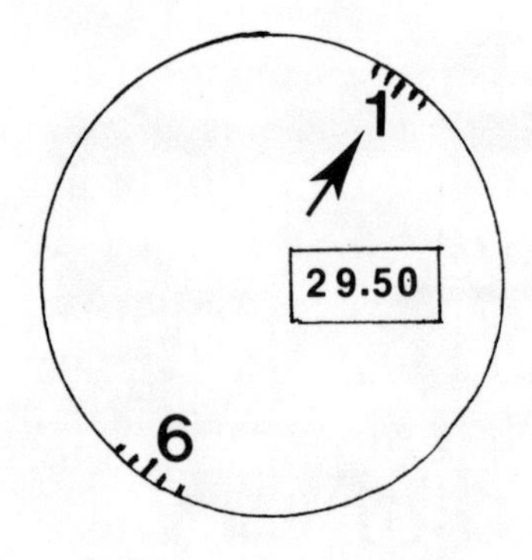

At takeoff time, note reading in the setting window.

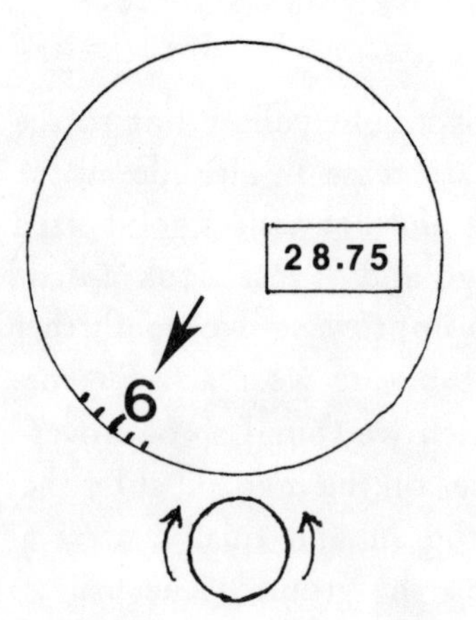

Set altitude back to arrival setting and note change in reading in setting window. It will not be exact due to temperature effects, etc. but it will give a fairly accurate estimate of any changes in barometric pressure during this period.

CHAPTER 26

PASSENGER PSYCHOLOGY

When you plan your trip, you will naturally have to consider your passengers interests and capabilities. Perhaps you can fly all day, every day and thrive on it. But your companions may not be able to. For one thing, you as a cruise director, may be engrossed in planning for and seeing new things along the way, but your passengers may not have the same emotional involvement that you do in seeing your creation unfold. Or, physically, they may just not be up to as much flying as you are. In many ways, riding in a small plane is more tiring than flying it. So, in planning or judging how much ground you can cover, the passengers' capacity for travel must be taken into account.

Concentrated Togetherness

The whole question of whom to take along for passengers or companions on an extended trip is always a rather delicate choice, but it is even more so for a wilderness or camping trip. Nothing can put a strain on a relationship with someone as much as travelling with them. A trip puts people together for longer continual periods of time than most any other social activity. In everyday life, a social relationship is tempered by outside forces; interpersonal contact is diluted by attentions to daily interests and activities. When not on a trip, people spending a full day together will have the time interspaced by such things as someone's leaving to run to the store, or by brief meetings with other friends or neighbors. The time together may be used for watching TV, cooking, playing cards or other distractions. On a trip, these diluting activities are diminished, and the people often have constant and uninterrupted togetherness without these psychological breathers. The contact is not only more constant but is also for a much longer time than most other social circumstances. Little differences or aggravations that may not be even noticed in ordinary situations, or that are noted and easily passed over, often become magnified by the prolonged and isolated contact of a trip. If the relationship is not one where each party can be completely free and open, the strain of having to "be nice" and to keep up a little social shield can soon become painful. It is a common observation that, after a trip, friendships are often changed; either considerably strengthened or noticeably weakened.

Lightplane travel often greatly intensifies these usual travel problems. The travel compartment is confining and isolated, and often noisy enough

to make full conversation difficult. There is not the opportunity which exists in many other travel conveyances for moving about or talking to other passengers. Stops are less frequent and harder to make than with an automobile. Private flying creates the need for more decisions than many other modes of travel. There are frequent options that have to be decided upon by the participants which may necessitate a continual compromise that can be wearing. The entire itinerary is constantly open for negotiation, as well as such things as when and where to stop for a break or for the night, how early to takeoff, when and where to eat, whether the weather is safe or not. All these decisions can put more strain on interpersonal relationships than most other trips do.

If you are a family unit, there will already be established patterns for working such things out. Old fishing buddies can do it with friendly sarcasm or horseplay. Others who have a relationship that is based upon frank and open expression of feelings will probably get along fine. In other cases, consider carefully before making a firm commitment for a long wilderness or flying camping trip. If the relationship with the potential "companion" is one in which you feel that you can't act completely naturally and have to "put on a front," unless you can overcome this barrier early in the trip, you may find that the confinement and isolation of the trip can be a psychological disaster.

Solo Trips

As mentioned in the Foreword, many of my trips are made alone with my dog. With a very busy professional, family and social life, I am almost constantly with people and very seldom have much time when I can do exactly as I want. These trips make wonderful chances to do this, and I bask in the enjoyment of doing whatever I want without having to compromise or even think about anyone else's wishes. This does not mean that these trips are conducted in a social vacuum; there are many chances for interesting social contacts on a solo trip. As Steinbeck describes in his

The author with Indian friends on a fishing trip up the Finlay River in British Columbia. One of the greatest satisfactions of a wilderness trip in your own plane is getting away from the usual tourist routes and seeing people and places that just can't be reached in any other way. Starting at a small strip already on the very fringes of civilization, these Indians took me in their crude wooden boats 25 miles up the river through spectacular mountain country which very few white men have ever seen. One could walk in any direction for three days from this spot and not reach any road. (The fishing was fabulous too!)

book *Travels With Charley*, a person travelling alone will often fit into a local social situation in a way that two or more people will not. Alone, your social energies are all turned outward toward the new acquaintances, whereas with others along, these energies are flowing within the group and not as readily available for new contacts.

So pick your companions with care, and if you wish, don't hesitate to do it alone. If you go solo, you are joining a select group which includes Thoreau, Lindberg, Amelia Earhart and Alan Shepherd.

Keeping Passengers Happy

Any passengers who have fears or uncertainties about flying may find these are increased in wilderness flying. The absence of any signs of civilization on the ground, the rugged terrain and the small strips can increase fearsome feelings. The thoughtful pilot can help minimize these problems. First of all, always fly as gently and smoothly as possible. Maneuvers that are commonplace for pilots can seem very scarey to non-fliers. Pilots often forget that just being in a lightplane is a big experience for the non-flier and it's easy to overload the novice with banks and dives. Secondly, try if possible to involve passengers in flying procedures. My wife became much happier about flying in Baja after she learned to use Senterfitt's book of airstrip photos and to search for and identify each one as it went by. Map reading, navigation checking and radio work can effectively take attentions away from things that may seem scarey to the novice. As I found out early in my surgical training, the responsibility of having to do something is the best antidote for squeamish feelings. In flying this can apply both to fearsomeness and to airsickness. All these things apply to flying anywhere, of course, but may be accentuated in wilderness flying.

Try to keep passengers as physically comfortable as possible. Back seats are sometimes less comfortable than the front ones, (it's a rare pilot who has ever ridden in his own plane's rear seats), and if passengers are crowded by luggage and gear the trip will be less happy for them.

CHAPTER 27

FLIGHT PLANS. HERE AND ACROSS THE BORDERS

CANADA • MEXICO • THE BAHAMAS
ROUND ROBINS CAN BE DANGEROUS

Flight plans are recommended in the United States, sometimes required in Canada and the Bahamas and automatic in Mexico (although only a formality there due to the lack of any organized search and rescue procedures).

CANADA

In Canada flight plans work exactly as in the U. S., except that the forms are a little longer and, if filed on the ground, have to be signed by the pilot. The Canadians seem to take them a little more seriously than we do at home, but this is probably just psychological because of the more sparcely settled country up there. Technically, they are legally required in Canada in sparcely settled areas if you are 25 miles away from the airport of departure, although they can be given to a "responsible party" instead of ATC, if the responsible party agrees to notify ATC if you are overdue. Many flights up there originate in isolated spots without communications by phone or radio; so some flights have to start without one. As at home, ground filing in person or by telephone is preferred, but they can be air-filed by calling the nearest Aeradio Station on 122.2 Carrying one of their flight plan forms will speed the radio filing. Closing them is strictly required and, as at home, can be done from the air, by phone or in person.

MEXICO

Mexican flight plans are a different story. The flight plan (which says *Plan de Vuelo* at the top) is a form that is filled out *for you* at each landing where there is a Federal Control Tower. It's automatic. You just answer the questions and sign it. English can always be understood but should be spoken slowly and clearly. If the next stop will be at another place with a Federal Tower, give that as the destination. You will be expected to show up there within a reasonable time. If you will be landing instead at a strip without a Federal Tower, give that as your destination. You can fly from non-tower strip to non-tower strip without filing any flight plans. Whenever you land at a Federal Towered airport, give your last non-tower strip as your point of departure and you will not be expected to have a flight plan from there. It is wise to always have your last made out flight plan with you. There is no apparent consistency or logic to what is sometimes wanted from you, and if you can show that you had a flight plan made out at the last towered strip, it often makes for happier relations. In some middle sized places, local authorities will have their own forms for landing permit or police forms. Don't worry about having to know about them. You will always be told about them (and about the small landing fee that some places charge). Always be sure that you have all forms given to you at the border with you when you are in the plane. Every so often, you will land at some very isolated strip and be met by heavily armed "Federales" (national police) who are apparently on the prowl for smugglers and drug runners. If you cannot promptly produce all proper papers you may have some heavy explaining to do, and their English is probably worse than your Spanish, especially when a "gun machine" is indolently pointed at your middle!

THE BAHAMAS

The Bahamas are casual about requiring flight plans. Nassau seems to be the only place there that really insists on your filing one. They are quite conscientious about following up on them and on having you close them upon arrival. You can almost always close one by air while at altitude before landing, but, as in Canada and Alaska, many spots have no on the ground facilities for closing or for filing one before take off. For such overwater flying you will probably want to have one. Nassau radio (on 124.2) covers most of the area well with several peripherals and you can air file a flight plan before you have climbed very high in most of the area. (Miami radio also has peripherals that cover most of the Bahamas well.)

ROUND ROBINS CAN BE DANGEROUS

For a direct flight to be completed the same day an American, Canadian or Bahamanian flight plan (augmented if necessary by position reports) adds greatly to safety. If it is not closed on time, these countries have a massive search and rescue organization that will start inquiring and eventually go looking for you. It is a wonderful free insurance policy, valuable anywhere but especially so in wilderness flying.

If a flight is to terminate at some location where radio or telephone closure of a flight plan is not possible, an American "round robin" or a Canadian flight notification (filed on a green form instead of the usual Canadian pink form) enables a delay of several days or more before the plan needs to be closed. In the contiguous United States these are used mostly for flights into Mexico, because there are few spots in the U. S. where there is no communication. In Canada and Alaska they are used more frequently because of the prevalence of isolated destinations. This of course, removes a lot of the protection, particularly for the outgoing flight, because there will be no search and rescue efforts until the expiration of the time the return flight is due. But it is still better protection than nothing.

The round robin or flight note does add an element of danger that is often overlooked. In questionable weather conditions, the "go no-go" decision often becomes weighted heavily on the "go" side because of the necessity of conforming to the timing of the flight plan. The authorities may have no way of knowing of the local weather problem and will be expecting scheduled closure. This has made many a pilot take off in conditions he otherwise would not have. The Canadians allow partially for it by routinely allowing 24 hours after the expiration of a flight note unless you request otherwise, and the same effect can be had by allowing an extra day in the timing of a round robin. (This, of course, also delays the start of any help this much longer.)

In this situation the procedure recommended to me by a Canadian Aeradio Station is to stay on the ground and call to "any plane copying" on 121.5. Do this when you hear or see a commercial jet overhead and if none show up for several hours start calling briefly every ten minutes. All commercial jets monitor 121.5 and most of the land areas of the continent are within radio range of a jet several times a day. If you reach one, ask him to relay your status to the facility he is in contact with. (If he can confirm back to you this message passage, it will make you feel easier.)

Once, when I was weathered in at a small emergency strip in the Yukon Territory this procedure was not working for me. Suddenly a large blue hole appeared overhead, undoubtedly the only one for at least 50 miles. I watched it for some time and it seemed about to stay around for awhile. Nervously watching the hole, I took off, climbed as quickly as possible to 9000 feet from where I could contact an Aeradio station, and extend my flight note and scooted back down through the hole. I am always very leery of "sucker holes," especially in such very mountainous country, but this experience demonstrates the danger-ignoring pressures that a flight plan can generate.

CHAPTER 28

BORDER CROSSINGS SIMPLIFIED

UNITED STATES • CANADA • MEXICO
ALASKA • THE BAHAMAS

For crossing an international border by private plane, many official publications as well as other flying literature often list quite awesome-sounding requirements that seem very complicated and confusing. The AOPA booklet on customs, for example, has 63 pages on U.S. Customs alone. After wading through such material, many pilots are apt to decide to avoid all the red tape and stay in the U.S. ("Maybe that little lake upstate isn't so bad after all.") But to be fair, the AOPA booklet lists data on nearly every possible situation and location, and in reality the procedures as they are actually practiced are much less complicated. The border officials in nearly all countries apparently do not read all of their governments' custom bulletins, or perhaps it just makes their work easier to keep things much simpler than the official documents direct.

The borders we are concerned with are between the United States and Mexico, the United States and Canada, the United States and the Bahamas and between Canada and Alaska.

FROM THE UNITED STATES INTO MEXICO

To cross from the United States into Mexico, file a VFR flight plan as usual with a U.S. FSS giving your foreign port of entry as the destination airport. No stops or inspections are needed in the U.S. Your first landing in Mexico must be at an "airport of entry," although it does not have to be the one closest to the border. (For example, you can fly from anywhere in California non-stop to La Paz, B.C. or Guaymas, Sonora, and clear Mexican customs there – if you don't stop anywhere in Mexico before this.) Close your U.S. flight plan as usual when you cross the border. At the appropriate time, call the Mexican airport tower just as you would any tower at home (but perhaps talk a little slower) and receive landing instructions in English using the same terminology and procedures as at home. (See also page 205.) After landing you will be directed where to go and told exactly what to do. They will make out several forms for you to sign, but they will do it all; you don't have to know anything about them. (It *is* wise to check and make sure that they have understood and put down the proper data for you.) If you don't understand anything or have any questions, ask them slowly and they will be glad to explain. They spend most of their time clearing American fliers and they are used to it. When they give you your papers, you are finished and can continue your flying exactly as you would at home.

That's all there is to it, so forget all the complicated requirements you may read about. *All you have to remember is the frequency of the Mexican tower!* (Some Mexican airports are hard to spot from the air, they tend to blend into the surrounding scenery, so the frequency of the VOR, which most Mexican airports of entry have, might be a help too.)

FROM THE UNITED STATES (OR ALASKA) INTO CANADA

For crossing from the United States into Canada the procedure is exactly the same except that there are fewer forms and when you contact the Canadian tower you should remind them that you need customs.

FROM THE UNITED STATES INTO THE BAHAMAS

Again, it's almost exactly the same procedure as entering Mexico and Canada. The Bahamians may be a little casual about it all and you should double check with them to be sure that all the required forms have been made out.

FROM MEXICO INTO THE UNITED STATES

When leaving Mexico for the United States, you must stop at a Mexican "airport of entry" and tell them you are leaving Mexico. They will, again, do all the proper forms for you. Again, it doesn't have to be the airport of entry closest to the border if you don't land again in Mexico after leaving it. While still in Mexico, radio the nearest FSS at least 15 minutes prior to your border crossing and tell them you are returning from Mexico and where (and when) you will be landing in the United States for U.S. Customs clearance. The FSS will take a short flight plan from you, and notify customs that you are coming. You will have to land at a U.S. airport of entry. Make calls to the tower or unicom as for any landing, and tell them you are arriving for customs. They will direct you from there and help you fill out the proper forms.

FROM CANADA INTO THE UNITED STATES OR ALASKA

Leaving Canada for the United States or Alaska is exactly the same, except that the Canadians will often give you a combined form when you enter Canada so that you don't have to stop for inspection on the Canadian side of the border when leaving.

FROM THE BAHAMAS INTO THE UNITED STATES

Again, it's exactly the same procedure as for leaving Mexico.

REQUIRED DOCUMENTS

To enter Mexico you will need a *passport or birth certificate* (any other type of identification usually will not do it). If you can conveniently do so before leaving home, a Mexican tourist permit obtained from a Mexican consulate's office or from some AAA offices will often speed up your passage at the border, but you can usually get one at the airport if you have the passport or birth certificate. A signed and notarized permit from a parent of any minors, not your own, may be required. The Mexicans may want a *signed permit* from the plane owner to take it into Mexico, as well as the plane's *Ownership Certificate* and *Airworthiness Certificate* and your *pilot's license.* The Canadians will usually be satisfied with only your pilot's license and another piece of identification. The Bahamas will want a birth certificate or passport. The United States, Canada and the Bahamas may ask for a *rabies certificate* for any dogs or cats you have along.

These are all the documents that you need, and its really very simple and routine; except that what you need, you really *do* need, and without it you may, literally, be denied entry, especially into Mexico.

To emphasize how easy a border crossing is and how few things you really need to have and to know, this chapter has been kept short and simple. The next one will discuss further helpful ramifications.

CHAPTER 29

CUSTOMS' CUSTOMS. (SMUGGLING IS NOT FOR CHILDREN)

HOW MUCH WILL YOU BE CHECKED? • LOOK OUT IN FLORIDA • MEXICO • CANADA • THE BAHAMAS IT'S CHEAPER TO PAY UP

The preceding chapter was a bare outline of the procedure of crossing international borders by lightplane, emphasizing what a simple process it can be and how little it is really necessary to know and do. Now to look at some more detailed aspects of custom clearances and some other matters that are involved in taking a plane into a foreign country.

Potential Contraband

All countries frown on importation by private plane of items for sale or for other *business* purposes, so if you are found to have such items, or items which even appear to have such uses, you may have some fast explaining to do. At every border there is interest in any *medicines or drugs* that may be aboard. Don't cross any border with any marijuana; one leaf or seed can get you into serious trouble. Alcohol is regulated at most borders, and entering the U.S. many plant materials, all *fireworks* and certain *fruits* and *vegetables* may be confiscated. Canada restricts the bringing in of handguns and wants to know about hunting and survival guns you may have, but allows them to come in. (The book says handguns are not allowed and will be mailed home, but if you are in transit between the lower 48 states and Alaska they will usually seal them at the border. The intact seal must be shown when leaving the country.) Into Mexico all guns (except hunting guns with necessary official documents) are taboo with a capital "T." Canada, the Bahamas and the United States require a currently valid *rabies certificate* for animals (see page 143).

There are limits for importation of *shrimps* and *lobsters* that are sometimes enforced very strictly. If you plan to bring any of these back, check it out first because the list varies from time to time and place to place. If you look like authentic campers or fishermen, the inspection is usually quite cursory, but you can't count on it! Customs violations by private planes are taken much more seriously than those by automobiles. The relatively little that you may save by running in a little booze, etc. is not worth the possible consequences. Penalties for contraband goods via lightplane seem to be more severe than those for an automobile. A friend of mine had his plane confiscated (literally padlocked at the field) for several weeks because he was caught returning from Mexico with twice the allowed limit of lobsters.

HOW MUCH WILL YOU BE CHECKED?

As with any type of border crossing, the amount of inspection will vary from time to time. Sometimes you could cross with a plane full of illegal immigrants buried under a loose pile of heroin and no one would know or seem to care. On other days a small bottle of aspirin may be

found way inside a suitcase and become the object of intense scrutiny. On the average, Mexican, Canadian, and especially Bahamian border inspectors seem to be less uptight about the whole process than their American counterparts. During many border crossings into these countries, I have never had the experience, for example, that I have had at the American border of being accosted by an inspector prominently displaying a sidearm in a manner implying that he felt he might need to use it on me.

U. S. Customs has a deservedly bad reputation with Mexican, Canadian and many U. S. pilots. Don't feel paranoid if upon return to the United States, instead of getting a warm homecoming you are treated like a suspected smuggler. On a bad day this can be routine. The lightplane is one of the main supply routes for the American marijuana market.

The enforcers are so frustrated at their inability to stop it that they take it out in their attitude towards all private planes arriving from south of the border. It should be repeated again: Do not under any circumstances have any dope in the plane at the border. Sniffer dogs may on duty, and one leaf or seed can mean good bye birdie, if not good bye for you too for a long time.

LOOK OUT IN FLORIDA!

If you are used to clearing U. S. customs into California or along the Canadian border and having inspectors come out and look at your plane, the set-up in Florida will seem tougher for you. I found this out the hard way. I had taken my papers into the customs building where the inspector looked at them and said, "You're all set!" In all my previous customs experiences this would have meant that they had decided not to look at my plane that time. I got back into the plane and was starting the engine when two inspectors rushed up and began pounding on the door and clawing at the door handle. It seems that in Florida you are supposed to carry all your cargo into the inspection building for examination. There are no signs or placards, apparently you are just supposed to know this. It's really an ineffective procedure because you could take in only what you wanted inspected, as I observed that none of the aircraft were checked.

Customs Charges

There is no charge for these services during regular working hours, but if you arrive in the evening or on weekends there is a charge. A maximum of $25.00 per plane is theoretically prorated by the number of planes arriving during a working shift to cover the cost of the inspectors' overtime salaries for this period. You fill out a form and get a bill in about a month (usually it is around $8.00). Some places try to collect the $25.00 in cash and promise to send you a refund of the difference. Airplane (and boat) owners have complained bitterly about this charge, which is not levied on automobiles, but, so far, without success.

It is advisable to keep all your Mexican flight plans, (the forms that say *Plan de Vuelo* near the top), as well as all U. S. bills and receipts if you had weekend charges. Once, some border official telephoned me to say that there was no record in their office of my having cleared customs on a border crossing some months before. Fortunately, it had been on a Sunday and I still had my bill for the customs charge. The number on this document seemed to satisfy the caller; but I have always wondered what would have happened if it had been on a weekday and there had been no bill. A Mexican flight plan indicating an intent to land at the U. S. port of entry might help. Perhaps one should take down the badge number of the inspector at each crossing, but this is a chore. Somewhere in the files there presumably would be a tape of your landing calls to the U. S. tower and ground, requesting clearance to customs, which could be obtained if really needed. The main thing to remember is that clearing with customs is taken very seriously. Apparently outgoing border crossing flight plans are sometimes checked for a corresponding returning customs clearance (the apparent cause of my phone call).

A friend of mine once purchased a plane whose previous owner had, some time before, violated some border crossing regulation with it. During a later investigation, my friend, as the then listed owner of the plane, was contacted. He explained to the officials that he had purchased the plane *after* the incident (which his papers confirmed) and that he had no knowledge of it. The conversation concluded with his being told that he should be extra careful on future border crossings because he was now on a suspect list for extra scrutiny. My friend was quite amazed by it all, especially, as he put it, at "having bought a crime."

It is often tempting to omit customs clearance. Especially if it is late and you are tired, it would be so nice to keep your altitude and scoot for home. The customs airport seems so small and far away and you are a bare speck up there – but it's really not worth the risk. For a proven violation, fines are commonly in the $500 to $1000 magnitude, and a temporary loss of freedom for your plane and sometimes you, too, is possible.

These things are all rare and exceptional cases, but if you are going to do border crossings you should know what *could* happen. Almost always the U. S. border officials are actually very polite, competent and friendly. If they feel that you are cooperating with them, you will almost always have friendly, helpful service.

MEXICO

Mexican border personnel are also almost always polite and helpful. Any inspection of the plane is usually once over very lightly, if at all, but an occasional spot check can give you a hard time. They will often question cardboard cartons. It is alright to use these, but they may want to look inside to see that they are not full of merchandise for sale. When you check into Mexico, a friendly, relaxed and patient attitude, at least outwardly, will ease your way through their form filling and other red tape. Every so often an official at the desk will announce an apparent sudden change in policy that could really foul up your plans. You might be given back your birth certificate or passport with a flat announcement that they no longer make out tourist cards there; you will have to go back to some Mexican consulate office in the U. S., or you might be told that you must have some document that has never been asked for before and which you don't have with you. These announcements are often given with a finality that would make the subject seemed closed. But keep your cool, don't get visibly upset. Any attempts to prove them wrong and you right will result in further strengthening of their stand. If you can just wait a few minutes quietly and then can manage to put on a slightly humble attitude of apology for not knowing about their regulations and with a friendly smile of naive innocence ask if they would do it for you just this once practically always they will very soon relent and do it for you. For them, it is a game of international one-upmanship, which relieves the pressures of their work and their social position. If you resist them, they will prevail. (They can always find some obscure regulation that will back them up.) If you back down (in their eyes) by being a little humble and asking their help, they can go ahead and grant your request and still feel that they won.

Don't Try To Rush Them

Mexicans work only at their own speed, which is often actually much faster than it appears to be. Any attempts on your part to help out and speed things up, which might help in the States, will only disrupt their accustomed work patterns and cause them to slow down. A relaxed smile will get you through much faster than any expression of impatience.

I once watched an American high pressure business executive enter the briefing room of a large Mexican airport. There were several pilots ahead of him and things probably appeared to him to be hopelessly bogged down, although a more experienced eye could have seen the hidden efficiency of the process. He was told to sit down in the waiting room as the rest of us had been told and had done with a casual smile. He decided instead to give them some Gringo suggestions for improving their efficiency. Receiving an indifferent response, he finally ended up by pounding on the counter and demanding to see the manager. He was repeatedly told courteously but firmly to take the seat in the waiting room, which he finally had no choice but to do. Our papers were finished shortly thereafter, but I had other things to do in another part of the airport. When I came back a considerable time later, he was still waiting, apparently for the "manager" while all the rest, including some that had come in later than he had, had gone happily on their way.

Most Mexicans are naturally polite and friendly, but make no mistake about it, you can get into serious trouble in Mexico. Underneath the friendly facade of many Mexicans is a latent resentment towards the U. S. and its citizens. If you study objective accounts of the history of the relations between the two countries, this is easy to understand. (30) Over the years, we have repeatedly given them the dirty end of the stick. The general attitude on both sides of the Mexican border deteriorated considerably when a number of years ago, during the Nixon administration a "crackdown" on the smuggling of drugs into the U. S. by automobile was made. During an ongoing campaign for several months, there did not seem to be any intensification of the actual search for drugs, rather the attitude seemed to be to make it more difficult for all Americans to return from Mexico. Obviously deliberate slowdown techniques were applied at the customs booths. The traffic backed up for miles and it often took several hours of waiting in line to get back home. The word quickly got around and a great many Americans stayed home. A considerable segment of the Mexican economy is dependent on the American tourist and was really hurting. The Mexicans were naturally resentful of the American refusal to modify this really ineffective procedure and began to be tougher on Americans coming into their country. The Americans finally abandoned their useless harassment, but the Mexicans still remember it.

There seem to be increasing reports of harassment of American pilots and confiscation of American planes by Mexican officials. The actual percentage is very small, considering the vast number of American planes that are continually flying into and out of Mexico. Nearly all of these incidents seem to have been triggered by a definite violation of some Mexican law, and it really does not appear that Americans and their planes are being capriciously held without reason. Still, the problem is there and it is a potential hazard that should be understood and faced before deciding to go to Mexico. If you should get into trouble, you are liable to harsher legal consequences than you have come to expect in your own country. Mexican law is usually based on the French "Napoleonic Code" which assumes guilt until innocence is proved which is the

A VIGNETTE OF AMERICAN TREATMENT OF MEXICO

What we say . . .

Sign on American side of the border

And what we do . . .

American fence seen from Mexican side of the border

These pictures were taken the same day and not too far apart on each side of the California – Baja California border.

reverse of most of our own legal ground rules (unless we are dealing with governmental agencies).

It's Cheaper To Pay Up

If you do break a rule or violate a law or regulation (or if you are accused even unjustly of doing something wrong,) try if at all possible to handle it immediately on the spot and as much as possible on a one-to-one basis with the accusing official. The higher up the chain of command it goes, the harder it becomes to resolve it and the longer it will take. If you can't come to suitable terms with the accusing official, at least try to keep it confined to the local airport crew. Once it gets beyond this, and into the legal bureaucratic network, its solution can become protracted beyond belief. If you have been caught with your hand in the cookie jar, admit it, ruefully and penitently, and ask how you can atone for your sin. It will often take a payment on your part. This payola or "mortida" goes against the grain for most of us, but it is part of life down there, and if you are there you have to play by their rules. The originally requested amount can often be gently negotiated downwards, but when you have what seems like their final offer – pay it if you possibly can, quietly, and quickly leave. Whatever it costs, it's cheaper at this level than it will be if it goes higher, and the time and harassment saved will be well worth it. If you are innocent, but they seem to feel that they have a good case against you anyway, consider the alternatives carefully. In the eyes of many of their officials, you are a rich, arrogant, ill-mannered foreigner, and your chances of fair judgment are slim. Galling as it may be, a quick payment of mordita and a quick escape may be by far your best alternative. (It's sort of like getting an undeserved traffic ticket at home. Proving your innocence is often difficult, and even if you do, the time and expense involved is usually not worth it – and the stakes are much higher down there.)

Apart from involvement with non-flying crimes such as robbery or murder, legal problems involving pilots in Mexico come under two categories: accidents, and smuggling or transporting drugs, guns or aliens.

Accidents

Many Americans have had accidents with automobiles in Mexico causing personal injury or property damage to Mexicans who could ill afford it. So many of these Americans have fled back to the U. S. ignoring all pleas for restitution that the Mexicans have found it necessary to detain all those involved in an accident (by car or plane) until responsibility can be determined and restitution guaranteed. This has made the standard Mexican procedure in case of an accident to place everyone involved in jail and then try to sort out the facts later (in some cases very much later). Mexican jails are easy to get into and very hard to get out of. While a few do have enlightened penal reforms, most of them make our jails look like country clubs. This is a definite hazard that should be recognized in making your decision to fly to this otherwise very charming and interesting country. Your accident insurance may cover *you* in Mexico (be sure and find out before you go), but the Mexicans have learned the hard way that they can depend only on insurance issued by Mexican Insurance companies. So you should consider arranging for this too before your flight, because without Mexican coverage any slight incident can result in jail. (31)

Smuggling

The transporting of *Guns, Drugs or Aliens* by private plane in Mexico is a very serious offense and you should avoid the slightest suggestion of appearing to engage in this. Have no guns along, unless you are a hunter and have proper papers for all your guns through the U. S. office of a Mexican consulate. Drugs are a no-no as we have already seen. Your passengers had proper documents at the border or would not have been allowed to go with you. Be sure you know any Mexican you may give a ride to while down there.

Mexican rules require that you return from the country with the same passengers that you went in with. Otherwise they suspect that you are running a charter flight service which, just as at home, requires special permits and licenses. Often you can return with the *same number* of passengers that you entered with and any substitutions will not be noticed. But if you have a different number of companions when you leave the country than you did upon entry you may have some explaining to do. If you know this is going to happen ahead of time, tell them when you enter and it will usually be okay. The same problem may arise if you show up at a large airport inside Mexico and the numbers don't jibe. It's not a real serious offense and you can usually talk your way out of it, but it can cause a delay and a hassle.

Always have all your border documents with you whenever you are in your plane. At the most unexpected times and places you may be suddenly confronted by a band of heavily armed "federales" (Mexican

National Police) who may assume that you are up to no good if you can't immediately produce the proper documents.

CANADA

By contrast with their two southern neighbors, Canadian customs officials are gentle. They have an old world charm, a refinement inherited from the traditions of the British Empire. They are gentlemen; and they assume you are too, unless you give them reason to suspect otherwise. You are expected to be playing by the rules and if it turns out that you are not, they can be very firm and positive in their enforcement of the rules, but it will be with an attitude of slight dismay and regret that you have made it necessary for them to be unpleasant.

Unlike most other foreign countries, Canada does not require that a General Declaration or similar document be filled out for a visiting plane. It is all done with a very small slip of paper given you after a quick check, primarily for handguns or commercial cargo. In smaller places, the customs inspector will often be waiting for you standing beside the parking area, and business will be conducted from a briefcase. Often you can get "combined" inspection upon entry and will not have to clear customs again when leaving Canada. You will need only some positive form of identification such as a driver's license, etc. The ownership certificate and airworthiness certificate should be available but will not usually be asked for. (Incidentally, notice how attractive the Canadian planes are without the large numbers on their sides.)

One time I entered Canada from Alaska at Whitehorse, Y. T. and obtained an in-and-out certificate. The next day I flew down to Juneau and returned to Whitehorse that evening. The same customs inspector was on duty. He asked me, apparently very seriously, why I hadn't just kept going on my return to Canada and used the certificate he had given me the night before. Such a procedure would be considered very illegal in any other country, of course, and I had just assumed it would be in Canada too. To me, this illustrates the simple honesty of our northern neighbors.

THE BAHAMAS

Fliers to most of the West Indies will normally clear customs in and out of the United States with The Bahamas. The Bahamians seem to be quite relaxed about customs matters. For one thing, their country has no particular industries that demand import protection and also life in all ways is more relaxed down there. If you do not get the service you are accustomed to it is probably not from an intended lack of cooperation but rather just a continuation of the low-key performance they have been used to all their life. Most personnel seem to be basically good natured and if you act relaxed and friendly there should be no problems.

CHAPTER 30

OVERVIEWS OF SPECIFIC AREAS

LOWER 48 STATES • MEXICO • CANADA
ALASKA • THE CARIBBEAN • THE BAHAMAS

This chapter will be a very simplified outline of the flying and travel features of several areas which have potential places for wilderness flying and airplane camping. It is intended to possibly spark an interest in an area. If this happens, the appendix will indicate where further reading can be obtained for more detailed study. (Information about flight plans, border crossings and customs details for these areas are taken up in the preceding three chapters.)

THE LOWER 48 STATES

This is home for most of us and we know all the procedures and regulations for flying here. We have a good familiarity with flying in at least some parts of it and have a general knowledge of most all of it. This country contains many areas with spectacular scenic wonders that make good flying destinations. It is getting more and more settled and civilized and wild areas are getting harder and harder to find. Many can still be found by lightplane, however, and some specific suggestions will be found in Chapter 30.

The advantages of staying within our own country are less distance to travel, a general familiarity with the language, customs and currency, no border crossing requirements and fewer health problems.

MEXICO

Mexico is a favorite flying place because it is handy; it is easy to fly to and yet it gives the real excitement and flavor of a truly foreign country. In almost every possible way, the contrast between Mexico and the United States is greater than that between any other two adjoining countries. The difference is even very noticeable from the air.

It's Different

When you first enter Mexico, you really *know* that you are in a foreign country. Everything is different. The smells, the noises, the language, the foods, the architecture and colors of the buildings, the currency, the whole attitude toward life – it is all very unlike anything at home. Mexico is popular with Americans. The San Diego – Tijuana border alone, has more crossings than any other international border in the world; but despite this continual influx of outsiders, Mexico remains relatively unchanged and different.

Yet, Mexico consists of many different regions that are entirely unlike each other. There are marked contrasts between the bawdy carnival atmosphere of the border towns and the vast areas of rural farmlands; between the extravagant hotels and fleshpots of the large luxurious seaside resorts and the miles of arid and practically empty deserts; between the busy gleaming ultramodern metropolitan areas and the isolated primitive natives still living in caves; between rugged snowcapped mountain country and miles of deserted tropical beaches. Even the airports show marked contrasts between super long and wide jet ports with elegant facilities and the small rough rural strips. The people vary too, those with wealth of an extreme seldom seen at home, live practically side by side with those with a level of poverty seldom seen at home either. Businessmen in elegant suits rush by on city streets passing shuffling barefoot indians who are just barely out of the stone age.

In a land of such great contrasts, there is something to appeal to nearly everyone, and this is one of the main reasons for Mexico's great popularity with tourists.

Flying in Mexico can easily reach all these extremes. There are landing places at super luxury resorts; at comfortable but less elegant fishing resorts; at rough fishing camps with primitive accommodations; and others adjacent to miles of fine sandy warm beaches with no one around at all. You can land and shop at the bazaars of the border towns; or in fancy metropolitan stores and boutiques; or at substantial shopping centers in large agricultural towns that never see tourists; or in picturesque primitive native marketplaces. You can land and visit the desert, the ocean or the mountains. You can land beside some of the worlds largest archeological ruins, in some places fully restored to their original grandeur and in other places still covered with jungle and only partly discovered.

Mexican Airports

The port of entry airports and a few other large metropolitan ones have a long length which is not always apparent on landing, because of a concomitant width and a blending of the surrounding ground into the bare countryside.[(51)] A long landing is usually helpful. The novice to Mexico can often be spotted because he lands near the approach end and then spends five minutes taxiing in to the briefing area. If there is a Mexican DC 9 on final behind you, the tower will often request you to land long, otherwise you have to figure it out for yourself.

Even though there is usually not much traffic on these big Mexican airports, it often seems to arrive at the same time that you do. If you hear a discussion in rapid Spanish on the tower frequency, you can usually assume that a Mexican plane is approaching the control zone. Don't let the Mexican controllers' accents fool you – they are really sharp. But it is still good to keep your eyes open if you hear the Spanish dialogue, because, unlike at home, you cannot tell from the conversation what is happening. Mexican tower operators all speak excellent English, but it is best to speak much more slowly and distinctly than with controllers at home. It will actually be faster, as well as safer, that way. If you were talking to the controller in person, a fluent conversation would be no problem at all. But, conversation is always more difficult over the radio, and the radio enhances any language barriers. The Mexican controller understands this and is not sensitive about it. Don't feel that you may hurt his feelings by asking him to say again, several times if necessary.

If he has difficulty in understanding you, try rephrasing your transmission in different words. Often Americans can be heard talking to foreign towers in short, clipped phrases that would be hard for controllers to understand even at home, and it delays everyone while the foreign controller is trying to figure it out. The unfortunate Canary Islands disaster suggests the problems that can arise when either pilot or controller (and especially both) are not using a native language.

Try a U. S. ATIS

Many Mexican airports along the border do not have an ATIS, but there is often one at a nearby airport in the States. It sounds obvious to suggest using one of these to get advance notice of the Mexican conditions, but it is surprising how often pilots do not do this and take up a lot of air time getting terminal weather information from the Mexican controller, which is available from a nearby American ATIS.

Most of these large Mexican airports have no taxiways. There is usually not much traffic so the tower will direct you when to taxi up the runway for takeoff and a delay is rare. Intersectional departures usually have very adequate lengths of runway, but you have to request them from the tower.

But most Mexican airports are uncontrolled, and usually deserted. Always call, about three miles out, on both 122.8 and 122.9 and state slowly your location and your intentions. Somehow what little traffic there is seems to bunch up. You feel all alone and assume that you are, but so often the only other plane for that week will be right behind you.

Anyone contemplating flying to Mexico, should get Arnold Senterfitt's book for either Baja or the Mainland, as the need may be.(32) These books are the pilots' bibles for Mexico and are an absolute must for intelligent and enjoyable flying in that country. They are interesting reading, even if you don't end up flying there. His map of Baja is a classic and extremely helpful, and especially when it is used in conjunction with his book, which it is keyed to. Tape the folds with transparent tape before using it, because it will get much heavier use than the ordinary sectional. His map of the Mexican mainland is helpful too, but is a modified IFR type chart that is not as helpful for touring. For the mainland, back this map up with ONC charts(33). They are worthless for airport locations but do show terrain features with fairly good accuracy. If you don't have the Senterfitt book, you can get the tower frequency of your airport of entry (and its VOR frequency) from the nearest U. S. FSS and then at the first Mexican airport get the tower and VOR frequencies that you will need.

Mexico has fewer VOR's than the United States and a lot of your flying will depend on pilotage or the ADF. In Baja the coastlines always have distinctive irregularities that make pilotage easy. On the mainland it's not quite so easy and, especially in the northern part of the country, the sandy desert and adobe towns all blend together and finding your location and spotting airports can be difficult. Unless you are within VOR or ADF range of a known location be sure and watch the charts carefully.

Mexican Weather

Mexican weather reporting is less sophisticated than in the United States or Canada and has been the brunt of various jokes in the past. Quite accurate terminal weather reports are usually available at larger airports – for other large airports. In between you are often on your own. Actually Mexican weather is usually quite good and reports are less needed than in many other regions. Weather often seems to come through in definite frontal patterns with good weather in between, but especially in the summer, afternoon cumulus buildups with thunderstorms are common. A lot of Mexico seems to avoid the lingering unpredictable "crud" that is found in many other areas. Best weather for flying (and general travelling) is in the Fall. Summer weather is quite hot and often muggy in much of Mexico.

Mexico offers many excellent spots of isolated wilderness for the flier, some of which will be described in the next chapter. But much of the charm of Mexico is in its people and their culture and you will probably want to spend a good part of your time seeing them. The real value of your own plane in Mexico will usually be found, not in locating complete isolation, but in its ability to easily take you to many infrequently visited areas where the true Mexico can still be enjoyed, unspoiled by the hoards of visitors that are now found on the usual tourist circuits. But, the isolated strips are there, and, sometimes very spectacular ones.

Mexican Camping

Camping is very popular in Mexico at the isolated and at the smaller spots. If there is a town of any size near the strip, a problem can arise because of the groups of little urchins that frequently hang around the airstrips. They are there to, hopefully, get paid for carrying luggage, "cleaning" your plane or just "watching" it for you, (sometimes this almost resembles a juvenile protection racket). Their enthusiasm in showing you their cleaning prowess often necessitates almost physical restraint to keep them from scrubbing the abrasive dirt on your finish with a dirty rag. They look so appealing and needy that it is hard not to dole out money, but those who really should know say that, unless they have actually done something active for you, such as carrying luggage, it is best not to give them money, although candy or a small treat is all right. Many of them are happy, friendly little characters and really harmless, so you may enjoy visiting with them. But if you are camping they can be a problem. They are persistent, interested in watching everything that you do and,

with nothing else to do anyway, they can be a real nuisance if you want any peace and quiet or privacy. They are used to firm, harsh treatment and you may have to be tougher with them than you really want to be.

The few large commercial Mexican airports will have no urchins. Neither will it have an atmosphere that is conducive to camping, if indeed, it is allowed.

Be a Gracious Gringo

It is often hard for Americans to understand the slight resentment that Mexicans sometimes seem to have for them. Until recently, throughout history, our country has always been a winner; not only militarily but also economically, socially and educationally. By contrast, ever since the Spanish Conquest, Mexico has consistently been a loser. Any time that a group of poverty stricken losers live right next to a group of wealthy winners there are bound to be feelings of resentment. Besides, over the years, many of Mexico's losing episodes have been due to the United States taking unfair advantage of the situation.[30] All these feelings are enhanced by the occasional "Ugly American" who appears on the Mexican travel scene.

Often the Ugly American is obvious, but at other times this attitude may be hidden. At the Baja "Road" Races in which Americans raced off-the-road vehicles over very rough Mexican back roads, the Mexican government apparently felt that the economic advantages overcame the obvious disadvantages of the racing cars tearing up the already very poor roads and roaring through small villages at speeds up to 80 MPH. The meetings between the Americans and Mexicans was outwardly very friendly and goodwill between the countries was toasted. Yet, one time I borrowed a Mexican truck to drive over to a nearby check-point. As each racer approached I always got far off the road and well in advance. But several yelled loudly at me, "Get off the road you ----- ----- Mexicans!"

All this is not your fault or mine, but when we go to Mexico it is something that we have to understand and to face. We are automatically assumed to be "Gringos" until individual Mexicans begin to know us better personally.

Even so, Americans can almost always get along and have no overt problems by using their *usual* manners and cultural habits. Many Americans do this all the time, enjoy the foreign flavor and have a wonderfully unique vacation. However, as part of our theorem of getting away from civilization by lightplane, we suggest that an even more interesting and enjoyable vacation can be made in Mexico by using a plane to get away from the usual heavily travelled tourist circuits and with only a slight modification of a few of our usual social habits get a little closer to the Mexican people and have a much more satisfying experience.

It is always somewhat of a shock when the travelling American first learns that our normal manners and habits seem a little crude and harsh to people of many other cultures. The average Mexican has a gentleness and politeness, an old world charm that we do not find at home. With only a little extra time and effort you can make yourself seem much more gracious and friendly to the Mexicans and get in return a much more gracious and friendly reception. Mexicans, among others, always use a word such as Senor or Senora at the end of a question or other direct statement to someone. In our culture it would sound strange and affected to say, "Isn't this a good morning, Misses?" But to the Mexican, to say (as we do), only "Isn't this a good morning?" sounds brisk and unfriendly, perhaps as if we were to say, "Hey, you, ain't it a good morning!" On my first trip to Mexico City, (always referred to as "Mexico" by Mexicans), we had ordered dinner at a nice restaurant, frequented by local inhabitants rather than tourists. The waiter approached with two plates and asked who had ordered the fish. I nodded toward my wife and said, "She did." As the waiter bent over to serve the plates he said very very softly and with only the barest suggestion of correction, "Por la Senora." Sometimes these are very subtle differences, but they can make your real acceptance much greater, and if you are away from the tourist centers your visit will be more worthwhile. In a social situation, as opposed to say in a store, it is proper in Mexico to begin a visit with a little polite conversation, before making a request or coming to the point of your visit. In Mexico, a person approaching a group of people whom he knows is expected to greet them first. To come up to a group and say nothing, expecting them to speak first, is considered impolite. Mexicans, incidentally, consider themselves "Americans." You are a "North American" (Norte Americano). If you remember this and respect their sensitivity about it, it is another of those little things that will make you seem much more gracious and friendly in their eyes, and in return your acceptance and treatment will be more gracious.

Language

English is understood, at least adequately, nearly everywhere, and even where it isn't, no real problems seem to arise. But unless you will be sticking to the larger tourist centers, (and you will be missing most of the fun if you do!), a small Spanish-English dictionary is indispensable. It is often hard to find a good one. Many seem to have been written by a Spanish Professor in some ivory tower and do not have words for ordinary everyday use. It should be small enough to carry easily, but large enough to have most words you may need. The best one we have found, after considerable research is listed in the Appendix [34] . If you can speak a little Spanish, don't be afraid to try it out. Mexicans are usually flattered by any attempts to use their language and will try to help you out. Some of them will try to get you to help them with some English words they want to learn. The only problem is that if you say something in passable Spanish, you will probably get as a response a torrent of fluent Spanish much faster than you can understand. If you smile and indicate your lack of understanding, it becomes a little joke and they will repeat in slower Spanish or in English.

Crime

The several revolutions in Mexico's history have given the impression that political leaders have a short life expectancy there and the exaggerated stories of Pancho Villa's really insignificant escapades of 55 years ago still help to portray Mexico as an unruly place with banditos constantly harassing tourists and political unrest imperiling foreign visitors.

For some reason there is a strong tendency for the American "media" to exaggerate any crimes in Mexico involving American tourists. Actually, as the Mexican tourist officials try to point out as tactfully as possible, there is probably no place in all of Mexico where crime is anywhere near as rampant as in New York's Central Park or in many major American cities. Still the legend seems to somehow persist and many Americans avoid going to Mexico, where they are actually much safer than at home.

In the past most Mexicans have been characterized by an almost absurd honesty, but like all the rest of the world this has been changing somewhat and thefts and robberies do occur especially in the major tourist areas, where they would not have in years past. Still it is all relative and today Mexico is actually much safer than many places in the United States.

Free Rides

If you become accepted by the local people it can add greatly to the satisfaction of your trip, especially in smaller less frequented places that your plane can so easily take you to. When this happens, you may find that requests for plane rides can become a nuisance. You will have to decide your own policy and stick to it. Often a plane ride seems like a good way to repay a favor that may have been done to you, but once you start, you may land and find that a whole slew of relatives and neighbors want a ride too, and, if your gas holds out, you can spend the rest of your vacation running a free barnstorming service. In several such situations, I have found that the best way out is to take off the top half of the cowling (which is quick and easy with my plane), and announce

flatly that the plane is "seek." An uncowled plane really does look sick to the non-flying eye and this really works. In case of necessity it is easy to "feex it" by fiddling a little with a screwdriver and then put the cowl back on. Be careful with the cowl top if you take it off. It easily blows across the ground in a light breeze and is easily scratched this way or by being inadvertently stepped on.

Mexican Aviation Fuel

The quality of Mexican Avgas is often questioned by some pilots who fly there. For ten years I have been using fuel from every conceivable type of dispenser in Mexico and never have had even one sputter from it. But I know of other pilots who have had serious troubles from either water or dirt particles in Mexican gas. There are three general types of fuel supply facilities in Mexico. The first is the large NACOA distributors (Nacional de Combustibles de Aviacion, S. A.) at the larger airports that have underground tanks or large trucks. They service commercial jets as well as general aviation and seem in all respects to be identical to what we have at large airports at home. Just ask for eighty ("ochenta") or "a hundred" ("cien") and it will be taken care of. If you feel better about it, they are glad to let you inspect the tank and cap afterwards. If there is a jet on the ground, it can delay the gas service for ten to thirty minutes.

I always order the gas before going in to get the paper work done; the gas is usually finished first. At border cities the gas often comes from the United States because it is dyed for octane identification which Mexican-made gas is not. This fuel supply seems to be in every way completely satisfactory and the price is substantially less than in the U. S. The pumps do not always seem completely reliable, but any shortage will be much less than the price savings. It is delivered in liters, so a little conversion table for liters and gallons is handy (39) . If you can, it is best to have the tanks completely filled, then you know how much you have for sure and are not dependent on their pump measurements or your metric conversion. The gas boys are very friendly and reliable. A tip is always appreciated because their salaries are low but is not really expected or necessary. BankAmericards and Master Charge (but not regular oil company credit cards) are accepted. They are made out in pesos and converted at the time of billing.

The second category of fuel distributors are independent suppliers at moderate-sized airports. They usually have large underground or overhead tanks and pumps. Usually the pumps have filters. These are often obviously not cleaned very often and a chamois filter may be advisable. The chamois will absorb a moderate amount of water and filter out dirt particles, and one should be used whenever there is any question about water or dirt in the fuel. If one is needed, it usually will have been asked for enough in the past that the dealer will have an adequate one. Inspect it yourself for holes, remove any left over dirt and be sure that the end of the funnel will not strike the bottom of your gas tank. (Holes have been made in gas tanks this way, and repairs are usually a long way off.) Advice is often given to carry your own chamois, but they are large, dirty, smelly and make unpleasant cargo in a small plane. The service at these smaller airports will vary, just as at home, and the price will usually be considerably higher than at a NACOA station. Occasionally an oil company credit card will be accepted, but usually cash is necessary.

The third category is the fuel available at smaller places, and it is here that the supply should be really questioned and inspected. It is usually stored and delivered from 55 gallon drums which are often rusty or sandy, inside. In some places the drums have to be floated ashore from a barge, and sometimes, somehow, they pick up a little salt water inside. This gas is delivered either through a hand pump on the top of a pipe which is put inside the drum, or sometimes it is poured or pumped into five gallon Jerry cans and poured into the plane from these. At least this way you know for sure how much is actually delivered, but it provides additional chances for dirt to get into it. In these places your direct observation and supervision is essential during the refuelling procedure and a chamois filter is essential. The price of this fuel will be high, usually considerably higher than at home, but the increase is really justified because of the effort and expense of transporting it there.

Climate and Clothing

Being farther south, the Mexican climate tends to be warmer than that of the United States, but the great variety of different regions naturally has a great variety of climates. General travel books on Mexico can give you a good idea of what to expect, but, especially with a lightplane, prepare for the occasional extreme not usually mentioned, especially the occasional short spell of cold weather even in the normally hot areas. Recommendations for appropriate clothing too, are given in guide books, but, again, the lightplane flier looking for the less "civilized" areas may need different clothing than the book suggests. There are places in Mexico where the black tie is appropriate, but the lightplane will also take you to areas where blue jeans are more suitable. Almost anywhere, a sport shirt and slacks for men and a simple dress for women will be comfortable and inconspicuous. Much of Mexico still frowns on slacks and shorts on women. Long hair on men was unacceptable for a long time, but they are catching up on this now.

Currency

For many years a Mexican peso was worth eight cents. In mid-1976 the peso was floated. This made a bonanza for American tourists whose dollars would suddenly buy nearly twice as much in Mexico. As long as Mexicans had transactions only within their own country (where prices were frozen) it made no difference to them. But if they had to buy foreign goods, naturally they suffered. They tended to resent the fact that a dollar now cost them nearly twice as many pesos, and at many places American money would not be accepted. There is no convenient place to exchange dollars for pesos at most Mexican airports, although, in a pinch, a taxi can be taken to a Mexican Bank or "Casa de Cambio" (Currency Exchange House). In normal times, dollars are good anywhere, but if there are any currency uncertainties it is wise to have along at least $50.00 worth of pesos, which can be obtained from most U. S. banks with some advance notice. The sign for peso is also $, which is sometimes a little confusing. At border towns or tourist centers prices listed in pesos will often say, after the amount "M. N." (moneda nacional). BankAmericards and Master Charge cards are good in many places in Mexico.

CANADA

Where Mexico is a land of contrast and variety, Canada is a land of similarity. It is just like home in most ways and often there is no feeling at all of being in a foreign country. Language, scenery, currency, foods, buildings, the general attitudes towards life, all the things that make Mexico so different are absent here. Canada is a foreign country and there are many differences but they are subtle differences and not very apparent to the casual traveller.

Canada has different regions and they correspond roughly to the same regions in our country. The East Coast has a very irregular coastline with cold water, farms and forests. There are older metropolitan areas in the East, and areas of vast plains in Canada's Midwest. The far Western regions match our own, and going north there is a region very similar to our Alaska. The most foreign-seeming thing is the French language which is dominant in some areas of Eastern Canada, and has been made a second official language in all of the country. Everything from flight forms to cans of food is printed in both languages, but outside of the eastern French areas only the English seems to be noticed. The controllers all speak English, of course (and presumably French too). They also understand American quite well, but an occasional place name can be a problem. Once I decided to stop at an airport spelled Courtenay on the map. The aeradio controller had never heard of an airport called "Cour-ten-aye" but soon asked if I could possibly mean "Kort-nie." (It turned out to be a charming place, by the way.) Some 80% of Canada is officially designated as a *sparcely settled area* which has some special requirements.[7]

For *ordinary* flying, there is not much to suggest other than to consider each area more or less similar to its counterpart in our country. For *wilderness* flying, however, there is much to suggest because Canada has some of the greatest wilderness regions left anywhere in the world. In fact, about the only generalization that could be made to differentiate Canada from the United States is that it is overall much less settled and populated. The whole country has much more open space and many more wild unsettled areas than ours. This increases in the north and reaches a maximum for wilderness recreational purposes in the untouched wilds of British Columbia and the Yukon and Northwest Territories.

Children in other countries can add a lot of flavor to your trips. These kids at Fort Ware, B. C., who spoke beautiful English, had never heard of Disneyland but they had as many real life "rides" of their own!

Canadian Airports

Canadian airports are very similar to those in the United States. The large ones are long and wide, well paved and lighted with excellent ground facilities. (Again you would think that you were home.) Their smaller general aviation strips are also like ours with similar facilities and the usual friendly pilots hanging around. The small and isolated Canadian strips in wild areas are generally good and, if anything, larger and better maintained than the isolated ones in other countries. In Mexico, a lot of primitive strips are not much more than a wide dirt road, but in Canada nearly all strips are wide and (at least in the summer) well maintained. Those that are not lighted or attended are usually at least marked at each end by large wooden panels painted with orange flourescent paint. There are also many non-mapped strips at ranches, sporting camps and the like that vary greatly in size and quality. Be careful of non-paved strips in late Spring because the surface may be very bad from the winter beating and the annual spring maintenance may not have been done yet. The maps and books often don't mention this, and it is often not apparent from the air, but the local pilots usually know about it.

The vast emptiness of so much of Canada has stimulated the development of an active and enthusiastic general aviation community. Even more than at home, there are flying clubs and enthusiastic fliers at small fields. Canadian pilots always seem glad to give any help or advice they can and hangar flying is a popular pastime. The Canadian Owners and Pilots Association (COPA) is an active and enthusiastic group.[35]

Canadian Air Controllers

Canadian air controllers are very competent and cooperative. They are used to working with pilots who take their flying seriously. There is often a camaraderie between pilots and controllers that is good to hear. Most flight plans air-filed in Canada follow the usual routine, but I have overheard this kind more than once:

"Canada Radio. This is CYE (Canadian planes have letters rather than numbers).on twenty two decimal two."

"Hi, Sam!"

"Oh, is that you George? Hi! Say, I left at 35 past the hour for the lake as usual. I'll be back tomorrow at 1400."

"Okay, Sam. Got it. We'll be watching for you 1400 tomorrow. Have a good flight!"

"Thanks, George!"

Probably very illegal, but also very effective. (I have been wondering lately if our FAA system isn't now sophisticated enough to use quick computer codes for the tedious, repetitious and time consuming data about plane and pilot presently used.)

But Canadian controllers can be formal and exact when the need arrives. I once overheard an Aerostar approaching Prince George with passengers and a malfunctioning landing gear. A planned wheels-up landing was arranged. A separate frequency was assigned to the project and the pilot talked to his company office, other Aerostar pilots, and to the very helpful controllers who discussed the surface conditions and crosswind components of the various runways. Ground personnel in trucks measured the height of the grass beside the runways and indicated where irregular spots were. Crash equipment was strategically placed and it all went with the finesse of a movie episode. He finally decided on the grass and made it without incident except for some grass stains on the belly.

"122 decimal 2"

The magic frequency in Canada (as in most of the world) is 122.2 It is used for most everything except towers, ground control and VOR's. If you don't have the frequency of a tower, you can call an Aeradio station, ("Kamloops Radio," for example) on 122.2 and ask for it. You could fly all over Canada remembering only 122.2 and get the necessary tower and VOR frequencies from Aeradio stations. It's not the greatest way to fly, but it would work. Most radio procedures are exactly the same as in the United States. Major differences in Canada are the relatively fewer VOR's, an occasional remaining "A-N" type radio beacon, and often lots of space in between. It's very easy to get lost, and *really* lost up there because of the lack of Navaids in many areas which have few towns, roads or train tracks guide you plus the confusion of many hills, rivers and valleys that all seem to look the same, especially if there is any weather around. Pilotage seems to come naturally for me. (I often wish other aspects of my flying were as good as my navigation.) But I have been lost in Canada, and more than once. One time near the U. S. border I was flying in adequate VFR weather trying to fly around some scattered showers of heavy rain ahead. Soon all the mountains and valleys began to look the same and I really didn't know within 50 miles where I was. The mountains prevented any help from VOR's or even the ADF. Fortunately, I had plenty of gas, so I headed south figuring I would cross the border and eventually pick up some radio aids. Suddenly a good airport showed up under me and I landed on it with the unique feeling of not only not knowing what airport it was, but not even *what country* it was in!

(It was Midway, B. C., ¼ mile from the U. S. border.) Most oldtime pilots in Canada have many tales of getting lost. One oldtimer gave me a helpful hint when he said he always used his thumb as a guide. When ever he had flown without knowing *for sure* exactly where he was for a distance equal to his thumbwidth on the map, he always made a 180 and went back to the last spot that he did know for sure where he was. If your weight capacity will allow it, always top off the tanks whenever you can. If you should become lost in this country, extra fuel is a priceless help.

ADF

An ADF is very useful in Canada because there are often towns with broadcast stations in the otherwise barren radio space.

Canadian Weather Reporting

Pireps are depended upon because of the large areas between weather reporting stations that can have unknown weather of their own. Each year more and more smaller places are beginning to send in regular weather reports several times a day. (This is helpful for flying, but it is another indication of how civilization is advancing and the true wilderness is shrinking. Quite a few places where even just a year or so ago you would have been completely on your own to judge weather now have regular weather reporting.)

At airports there are two distinct agencies. One is the Aeradio station which handles flight plans, airport advisories and clearances. In a separate room will be the Weather Office. Both facilities are always very capable and cooperative.

Canadian Maps and Charts

Probably the best charts for Canada are the WAC (World Aeronautical Charts) series. They are very similar in format to the regular U. S. sectionals and give the same information. They are published by both the United States and Canada. The identical topographic map is overprinted differently by each country with minor variations but each is very adequate. The scale is 1:1,000,000, so if you have been using U. S. Sectionals, (which are 1:500,000), you will suddenly seem to be covering the map only half as fast. For this open country the smaller scale is very adequate. A very big help in Canadian flying is the "VFR Chart Supplement" which is revised every seven weeks and gives information about airport and navigational frequencies, etc., some of which is often hard to get in any other way. Usually the current one is posted at major aeradio stations, but it is often hard to buy your own there. You can order one by mail (order it well in advance of your trip).[36] Canada is also covered by 117 "Aeronautical Charts" similar to the U. S. Sectionals, except that they are smaller in size and each one covers considerably less territory. Their use is often recommended but they show no more information than the WAC's, the same data is just printed on a larger scale. The WAC's are less bulky, cover all the same material and are considerably less expensive. Canada is also covered by two series of Topographical Maps which are helpful for hiking and camping away from airstrips. The flying maps can be purchased at larger Canadian FBO's or ordered in advance by mail. The same office can supply the Topo maps and a free index to them.[37]

A booklet on flying in Canada can be obtained free[38]. It contains a lot of information, some of which is useful. There are no real inaccuracies in it, but it is written to describe how the officials intend things to be done, and sometimes the actual practice in the field may be a little different, usually easier and simpler.

Canadian Aviation Fuel

The usual types of Avgas are available at all large and medium sized airports, from private dealers, in all ways similar to the set up in the United States. The small strips usually do not have fuel, although at some of them it will be brought in from a nearby town (at extra cost) on advance notice by telephone, or you can wait for it after you arrive. If possible, it is easier to gas up at the regular fuel facilities at larger places. In the sparcely settled areas always take on all the fuel you can safely carry, for it may be a long walk to the next gas station!

King Sized Gallons and Quarts

When you first fly in Canada, you may find it a happy surprise that your plane is suddenly getting about twenty percent better gas consumption. If you have not also noticed that the costs are likewise about twenty percent higher it may seem like a nice little bonanza; but it quickly collapses when you realize that it is due to the larger Canadian Imperial gallons.

If you forget about it, the error is on the safe side because you will have gotten more gas than you expected from the number of gallons put in. However, if you fly in Canada long enough to get used to the larger gallons, upon return to the U. S. you will be getting about twenty percent *less* gas per gallon than you have become used to.

If you are keeping accurate records, or if you are not filling your tanks completely, use a table to see how many "ordinary" gallons you are getting. Otherwise, you can just top it up every time and fly as you always do. In Canada, quarts of oil are king sized too, so be sure that you don't overfill without realizing it. [39]

Currency

For many years the American dollar was worth five to ten percent more than the Canadian one, yet no one in Canada seemed to notice this. For several years now the situation has been reversed, and now a conversion factor is usually charged when you pay with American money. Most major credit cards are good for Canadian gas. (The conversion factor is added when you are billed.)

Clothing

From a social standpoint, the proper dress in Canada would be the same as for the same situation in the United States. From a safety standpoint, however, much warmer clothing may be needed when flying in Canada. Especially in the northern sections, a sudden weather change may leave you stranded in very cold weather which at the same time makes flying conditions unsafe for leaving. Adequate warm weather gear is essential here, not only for comfort, but possibly for actual survival. (See pages 93-102.)

ALASKA

Whatever superlatives you may have heard about Alaska were probably understatements. Alaska is a land of extremes, if you like true unspoiled wilderness country, virtually untouched and unchanged by man. After you have explored and camped in Alaska, anywhere else will be a letdown; by contrast with your Alaskan experiences it will be sort of like exploring and camping in a city park. Naturally, a place as extensive as Alaska has areas of varied physical characteristics. There are flat plains of barren tundra which stretch for hundreds of isolated miles, contrasting with other tundra areas which cover rugged and spectacular mountains. Wild mountain streams, often laced with gold and teaming with fish, rush down into slow, majestic, tortuous rivers which meander past hundreds of miles of complete solitude. Forests of unequalled density cover other enormous areas. There are ocean beaches and dramatic fiords. Elsewhere are high snow capped peaks, glaciers and iceberg-filled bays. This country is the last refuge for a great many species of large wild animals.

Climate

Alaska's old image was of a land of eternal ice and snow inhabited only by Eskimos living in igloos while subsisting on whale blubber and by parka-clad trappers with sled dog teams. But, now, everyone knows of the other faces of Alaska and of its very pleasant summer weather. The winters are long and unbelievably harsh, much more so in the interior (Fairbanks) than along the coast (Anchorage and Juneau). Airplanes, along with other machinery are very hard to keep going in the consistently low minus temperatures. Various heaters, taking the oil and battery inside at night, and other irksome chores keep engines barely running. Winter flying conditions can be bad with severe storms, "white-outs" and other added flying hazards. The outside limits for any likely pleasurable flying travel are May through October. June through September is probably more realistic, and July and August are really the best. The mosquito density varies from year to year, but they tend to be much less numerous later in the summer. Considerable rain can be expected at any time, but usually it comes in definite short frontal passages. Otherwise the summer weather is sublime.

Tiana of McKinley

In most areas, the ground, except for the top few inches, is frozen all year ("permafrost"), but you would never know it unless you are an expert on the way the trees grow or unless you dig a few inches into the ground. This permafrost, along with the other weather extremes, necessitates ingenious techniques for the construction of homes, roads and other projects, but other than as something of possible interest, this will not effect you as a visitor.

Even in the summer, care must be taken to have adequate warm gear as described in Chapter 16.

The summer days are long. When I first went to Alaska where flying trips are proposed after dinner, I questioned flying in such country after dark until my hosts smilingly reminded me of the midnight twilight.

Alaska can be divided up or classified in many different ways. For flying purposes, there are two main areas, the main body of Alaska and the lower part, often called the Panhandle or Southeast Alaska. This lower part has very unique and potentially dangerous flying conditions because of the lack of roads, high mountains often extending right to the shore, many large ocean bays and a predilection for very poor weather. If you can safely get there, it is magnificent country, unlike anything else in Alaska (or most anywhere else). But the weather is notoriously unpredictable and noted for sudden change. It is easy to get caught there for a long spell of bad weather and IFR conditions are very risky for a small plane.

Aurora Borealis

The "northern lights" make an unbelievably magnificent natural light show. Even after all the superlative visual stimuli we are exposed to with our electronic marvels, nature's old display still seems very dramatic and breathtaking. These are largest and best in winter, but I have seen amazing ones as early as late August. They really are almost worth the whole trip to see.

The Pipeline

As the largest non-military project ever undertaken, the pipeline was bound to have far reaching effects even to an area as vast as Alaska. Fairbanks and Valdez used to be quiet, friendly little cities where it seemed as if everyone knew everyone else. They were literally inundated with construction workers of all sorts. The pipeline siphoned off so much of the normal working force with fabulously paying jobs that ordinary businesses and services suffered. After the pipeline is finished, most of these extra workers will leave for warmer climates, but some effects on the remaining population will remain. While this is of extreme importance to the Alaskan, it does not have too much effect on you, the flying visitor.

Crime

Normally, Alaska tends to be a peaceful place. The hometown atmosphere and the lack of slums, ghettoes and other crime-breeding conditions keep it relatively crimefree. The influx of so many outside pipeline workers as well as the large amounts of money and materials involved naturally changed this a lot and many problems have arisen. Still, overall it appears peaceful as compared to many places in the lower States. A recent series in the *Los Angeles Times* headlined "Fairbanks: Armed and Dangerous," mentioned its "notorious" Second Avenue and implied that there were dens of criminal activity there. Well, after long visits to Fairbanks and after observing people routinely leaving their homes and cabins unlocked and keys in cars and trucks overnight, after shopping on the "notorious" avenue and even walking there untroubled at night, I have concluded that many cities in the lower United States should be so "armed and dangerous!"

Flying Conditions

Many pilots think of flying to Alaska as a hazardous ordeal. Actually, with a few relatively simple preparations and, if done with adequate respect for weather, it can be very pleasant, relatively safe and extremely rewarding for the outdoor loving type.

Flying conditions, except for the mountains and sparcely settled country are the same as in the United States. Survival gear requirements (8) are about the only different legal requirement, but you will find that the general attitude towards flight plans, position reports and pireps are taken more seriously than in the lower states. But nearly all aspects of flying such as getting the weather, obtaining fuel and radio procedures are all exactly the same as at home. If anything, you will find a little more cooperation and help from controllers and other pilots because they are used to working in isolated areas and with more hazardous flying conditions. Flying in Alaska is more like it was in the lower 48 states some years ago. The pilot is still a member of an exclusive fraternity up there. Those that fly consistently in that country take their flying seriously and have learned the value of friendly cooperation. There are still relatively few of the "Sunday driver" type of pilot who seems to have proliferated in the lower states in the last decade or so.

Alaskan Airstrips

A lot of pilots have asked me about the conditions at airstrips in Alaska. Many seem to have the impression that flying to and in Alaska will mean having to use strips with very poor surfaces covered with large rocks. While you can find strips like this in Alaska, they are the exception rather than the rule. The usual airport in Alaska will seem exactly like the equivalent one at home. You can fly all the way up there, and most pilots do, using only long, wide paved strips with all the aids you are used to at home.

Beyond this it is up to you. You can explore a lot of isolated country using very adequate non-paved strips without any difficulty. The only difference is that because of the vast spaces with no ground transportation, the airplane is used to supply a great many villages, mines, sporting camps and ranches. There are, therefore, a great many more primitive type strips than there are in the rest of the country, and naturally some of these will be of rather poor quality. A lot of small mines or homesteaders have built strips in the best spots they could which often means that they are short, narrow and perhaps have hazardous approaches. For the owner, experienced with the strip, its surroundings, and local weather and wind quirks these are adequate, especially for his tundra-tired Super cub. So, you will see many that you will want to pass up, but there are still many that are safe and adequate, and you can fly over most of Alaska without needing to use really poor strips.

HOW TO REACH ALASKA BY LIGHTPLANE

Alaska seems to be much more of a challenge to the private pilot than the other areas in this book. Planning a trip to the Caribbean is quite cut-and-dried. Master the overwater bit and navigate well enough to hit an island and otherwise it's almost like flying anywhere else. Mexico is a little more of a challenge for most beginners but again it's really only an extension of home-type flying moved to a foreign country plus fewer navigation aids and perhaps some rougher strips. But Alaska! In the minds of many fliers this seems to be the supreme challenge. Fears of impenetrable mountains, the nearly total lack of civilization in many parts and the old stories of the hardships of gold rush times all blend to build up psychological barriers.

To be sure, flying to Alaska is not like flying in Kansas. There are things to be cautious about and we have attempted to outline many of them in this book. But Alaska is safely flown to by a constantly increasing number of pilots, most of whom wonder afterwards what all the fuss had been about.

So a little more direction may be needed for the would-be Alaska flier than for our other areas. Most fliers to Alaska follow the highway, (although there are some deviations from this which are popular.) To get to the beginning of the Alaska highway there are various satisfactory flying routes, depending on where you are starting from. These are adequately described in the AOPA Booklet on Alaska [40] and in *The Milepost.* [41]

Following a highway is often a good safe policy but it's easy to become too complacent about using highways for emergency landings, especially in wild areas. Many of them are very tortuous and rough. After you have seen some of them on the ground you will realize that many highways in timbered country cannot be reached by a plane because of the umbrellas of tree branches over them. Unless a highway looks awfully good from the air, keep an eye peeled for other nearby possiblities.

The principally used *deviations* from the Alaska highway are the route from Whitehorse to Dawson City which is good, although some may find the stretch beyond this into Alaska a little wilder than they want. The other principle deviation is "The Trench" which is a more or less continuous valley between parallel mountain ranges extending from Kalispel, Montana to Watson Lake, Y. T. I have flown over all of it and found a few borderline stretches in lower British Columbia. But the usually used section is from Prince George to Watson Lake and is, at least for me, adequate. (See page 238.)

The *coastal route* from Vancouver via Juneau to Anchorage covers long stretches of water, has notoriously poor weather and the underlying terrain is usually very inhospitable. For lightplanes its conditions are far below my minimums.

To examine any of these routes use the technique given on page 167 to inspect them on the map [42] and if you decide to try one keep in mind the suggestions on page 51.

Currency

Naturally, the currency is the same as in the rest of the United States, but there seems to be an exaggerated idea of how much higher prices are in Alaska. Prices for things that need labor or services are *much* higher. Restaurant prices and hotel rooms are probably nearly twice as high as the rest of the U. S. Aviation fuel runs 35-40% higher. Groceries in small "all night" type markets are about twice as high, but in the larger supermarkets most items run only about 50% higher. If you are going to eat in restaurants and stay in hotels, (rooms are very hard to get, incidentally), it can really strain your budget, but if you are camping, there will be only a moderate increase in the total cost of your trip.

THE CARIBBEAN

Flying in the Caribbean is a new experience, entirely different from flying anywhere else. Literally thousands of islands of all sizes, widely varying in geography and politics, lie in a sea of incredibly blue and clear warm water. It is one of those places that is impossible to describe adequately on paper. Almost everyone who has flown there returns full of enthusiasm for it and a strong desire to return again soon.

Like Mexico (as perhaps contrasted to most of Canada and all of Alaska) there is something for almost everyone. Historical buffs can revel at the site of Columbus' landing and visit some of the oldest continuously settled cities in this hemisphere. Nearby other beautiful areas contrast by being almost completely untouched by civilization. Very luxurious resorts abound where the visitor's every whim is catered to in full elegance. Then there are a multitude of less affluent resorts which vary in luxuriousness (and in expense) from super ones to simple beach cabins. Sailing, yachting, fishing and skin diving are superb. Genuine foreign shopping trips can be made to fine stores on islands that are parts of several European countries. Other islands are countries of their own, run by their own people. Camping spots are limited, compared to many other areas, but excellent ones do exist. (See page 243.) All this is wrapped in a mild tropical climate that is warm and relaxing.

Politics

Unfortunately, at the present time all is not harmonious in this paradise. The Caribbean washes the shores of land belonging to some seventeen different countries. These range from possessions of the United States and several major European countries to the communist stronghold of Cuba. In between are various smaller islands, many of which are countries in themselves which sometimes have quite definite political ideas of their own.

There are now a number of areas that are not pleasant or even safe for outsiders to visit. In the spring of 1977 I talked with a number of different pilots in Florida who had had long experience flying in the Caribbean and they all said that there were places that they used to really enjoy that were not safe to go to now. These pilots did not seem like the type to be easily intimidated and they also seemed to have solid evidence and examples to back up their opinions.

The problem appears to arise from two sources. The movement for independence which is strong on many of these islands has made the inhabitants of some of them want to exclude *all* outside influences, even tourists. In some areas the people are apparently encouraged by their governments to make it unpleasant and sometimes actually hazardous for travellers to visit there. The other source of danger is the inspired agitators whose political indoctrination makes them sincerely and fanatically believe that they are justified in doing in as many capitalists as they can get away with.

The situation is so volatile and unstable at the present time that no permanent guidelines can be given. Fliers must make their own evaluations of the situation at the time they are deciding to go.

THE BAHAMAS

For these reasons, specific details for the Caribbean will be given only for The Bahamas. This does not mean that The Bahamas are the only safe place to visit in the Caribbean at the present time, but the situation there is excellent for visitors and shows every promise of staying that way in the predictable future. Also, wherever one may fly in the West Indies, it is likely that border clearance both ways will be through The Bahamas.

Border Crossings

Details for flight plans, customs and radio frequencies for The Bahamas have already been given in preceeding chapters. They are easy and simple, much more so than some of the literature makes it sound. The Bahamas have many more airports of entry than most other countries and customs clearance on a quite informal basis can be obtained at many places, both inbound and outbound. Some of the Florida FBO's will fill out required Bahama forms for you before you leave the U. S. and save time on entry. Some of them can also give much detailed information about flying and the up to date tourist conditions for most of the West Indies. [43]

Flying Conditions

The Caribbean is a beautiful place to fly. The colours of the water are unbelievable and its clearness makes the bottom easy to see over large areas. The water has a warm friendly appearance that is somehow different from the somewhat harsh and hostile feeling that water seems to impart to the flier in most other areas. The weather is usually good for flying. Even the bad weather is often either isolated in small areas or has ceilings high enough to safely fly under. Between the larger islands are thousands of smaller islands, or "cays", varying in size from just a few rocky ridges protruding above the water on up to the larger islands. Radio coverage is good over the entire area *(See page 185.)* and, with proper equipment, flying is safe, easy and enjoyable. The Bahamas require a Coast Guard approved life jacket for each occupant and many fliers feel safer with an inflatable life raft aboard. Both of these can be rented on a daily basis from many Florida FBO's. [43]

Staniel Cay Airstrip in the Exuma Islands of The Bahamas. Typical of the good strips found in this area. This one serves two resorts and takes up a good part of the entire island. *(See page 243.)*

Bahamian Airports

The airplane is a major method of transporation in the West Indies and most airports are very good. Almost any island of any size has, or is near to, an airstrip. An excellent Bahamas Air Navigation Chart which will help your trip planning is available free. (44) Many strips are labeled "private," which has a little different meaning than in most other countries where private often means No Trespassing. In the Bahamas, private refers to strips not owned or operated by the government and, at most of them, visitors are welcome.

Camping

The Bahamas (and the rest of the West Indies) have good spots for camping. The excellent climate and isolated beaches are inviting to campers. Yet, camping is not too well accepted in many places. For one thing, many of the airstrips have been constructed by a nearby resort under very difficult and expensive conditions. The resort owners understandably want users of the strip to use their facilities to get a return on their investment and are naturally resentful of someone who uses their expensively built strip and brings his own accommodations and food. Secondly, campers seem to have a bad reputation in many areas here. It is hard to tell if the incidents causing this attitude are common or just extra well publicized, but, in any event, campers in many places are looked upon with skepticism. There are some very well located strips that are quite isolated and probably no one will even know you are camping there, but if there is anyone around it is wise to ask before camping. Some of the Florida FBO's often know which strips will currently allow camping. (43)

CHAPTER 31

DESCRIPTIONS OF SPECIFIC PLACES

LOWER 48 STATES • MEXICO • CANADA
ALASKA • THE BAHAMAS

This chapter describes a few sample places both for wilderness camping and for staying at isolated spots with accommodations. As mentioned in the foreword, the purpose of this book is not to describe specific places to go to, but rather to be an instructional manual with suggestions on how to find-your-own comfortable places. The places here are listed primarily to give examples of what to look for and what to expect in places that you may locate on your own.

CAUTION

With the two exceptions that are noted, I have personally landed at and inspected all of the locations listed here and have described how they appeared to me when I was there. However, wilderness places often have a very delicate natural balance that is easily upset and changed. They also do not have the sometimes stabilizing influence of civilization with such things as strip maintenance that help keep the status quo. Therefore anyone attempting to visit any of these places must anticipate that flying conditions may have changed for the worse since I was there and must make suitable inquiries on his own as to the present conditions and safety for his use.

LOWER 48 STATES

CAPISTRANO, CALIFORNIA *(Los Angeles Sectional)* I almost forgot to include this one because it is "home" for *N92523* and I think of it as a working place and not a vacation place, but it is actually in the center of one of the greatest resort areas in the world. Camping – not at the airstrip, it's okay with the management, but nixed by the local fuzz who are, fortunately for those of us who tie down there, quite protective. However, a State Park campground a half mile away is right on the beach with full facilities. (Write to Department of Parks and Recreation, P. O. Box 2390, Sacramento, California 95811 for full details, rates and instructions for the reservations which are usually necessary in season.) Nearby too, are famous surfing beaches *(San Onofre, Trestles)* and not too far away are famous commercial tourist attractions *(Lion Country Safari, Disneyland, Universal Movie Studios, Marineland, Sea World,* etc.). Nearby too, is the *Mission San Juan Capistrano* of the swallows fame. A local marina provides rental boats of various sizes for ocean sailing, and fishing boats go out for a half or whole day at reasonable rates. In season (November thru March) whales can be easily observed from the air just offshore. An interesting flight over the ocean for 24 miles enables one to land at an excellent airport on *Catalina Island* and tour this unusual island by minibus. About 45 minutes flight time away the *Anza Sky Trail* starts. This is a very unique "self guiding air trail" which gives pilots an unusual view of the geology, history and flora of the true desert country with the *Salton Sea* included. (Send $1.00 to the Sacramento address above for two booklets and a map for this unusual flying experience.) Excellent restaurants of all varieties are prevalent in the area. For all of these things supplementary ground transportation is needed, but car rental or taxis can be obtained through the flight office. This is obviously not a wilderness area, but it is an area with one of the greatest concentrations of interesting tourist facilities anywhere.

CAVANAUGH BAY, IDAHO *(Seattle Sectional)* A long wide grass strip right beside *Priest Lake* in beautiful forested mountain country. (Strip is not to be confused with the Priest Lake Strip which is not right on the lake.) Not true wilderness but a very pleasant woodsy strip with a good campground right beside the parking area. Small grocery store and pay phone close by. Quiet off season, it can be crowded in season. No winter use except for ski-planes.

CHAMBERLAIN, IDAHO *(Great Falls Sectional)* One of a number of U. S. Forestry Service strips in the *Idaho Primitive Area* and adjacent country. It was written up in the *AOPA Pilot* several years ago and has been almost too crowded ever since. Still it is a good pleasant place in wild country. The USFS maintains a summer station there which removes some of the true wilderness feel but the rangers are friendly and interesting and they have an emergency communication facility that might be useful. I have flown in from Hamilton and Missoula, Montana and McCall, Idaho. The latter is a shorter and safer approach but it is still not like Kansas! Even so, it is probably a good starting place for those who want to "get their feet wet" and experience nearly true wilderness and still not have complete isolation. Camping facilities are rustic: tables and USFS iron fire boxes. A large stream nearby adds interest. (For more details and colored pictures see the *AOPA Pilot* for June 1971, page 93.)

COLUMBIA, CALIFORNIA *(San Francisco Sectional)* Camping area under trees adjacent to west (dirt) runway. Tables and outhouse, no water or other facilities. Drinking water and restroom in flight office ¼ mile away. Camping area is quiet and private, but roads and houses are ½ mile away. This is a good spot for easy overnight camping and handy to air routes for many areas of the west, but after you have visited the restored gold rush town a short walk away with interesting shops and displays, you probably will not want to spend too much more time here.

CONDON, MONTANA *(Great Falls Sectional)* This is another not completely isolated place where you can get some of the feel of the wilderness without the privations of complete isolation. A State Forestry Service camp is adjacent to the strip. No one there will bother you, but it does offer emergency communications. No camping facilities but you can pull off and camp under the trees. A fairly busy road runs adjacent to the other side of the strip but there is not too much traffic and its noise is not bothersome. Grocery store 3/4 mile down the road (south). The strip is fairly short for its elevation and tall trees are near both ends but it can be reached without having to fly over dangerous country. Surface is fairly well kept turf (mowed grass). Could be a density altitude problem in hot weather because of the irregular surface and tall trees.

DUTCH JOHN, UTAH *(Salt Lake City Sectional)* Not true wilderness but a pleasant isolated strip adjacent to the *Flaming Gorge Recreational Area* (large lake behind a power dam). A good strip but approach can have trickey up and down drafts in certain weather. Camping without facilities beside an unused runway. Pleasant FBO operators with simple but competent facilities.

McKENZIE BRIDGE *(Klamath Falls Sectional)* One of my favorite spots in the lower United States. See page 175.

PORTHILL, IDAHO *(Great Falls Sectional)* This country strip is best known for its unusual border crossing custom arrangements. The grass strip is adjacent to the Highway Customs Offices of both the U. S. and Canada. Fliers going in either direction can land and taxi to a large field beside the highway and customs inspectors from either country as needed will come over from their booths on the highway and take care of

Photo by Betty Grindle

Porthill, Idaho *(Eckhart International Airport) showing both U. S. and Canadian customs areas and the bucolic setting.*

the formalities. There is a pay phone with which flight plans can be closed or filed with either Spokane Radio for the U. S. or with Cranbrook Radio for Canada. This, too, is not wilderness but there is a very large field far enough from the not too busy highway that it is a quiet and restful place to stop. The customs has a Unicom (122.8). The whole atmosphere here is more pleasant and relaxed than at many airport customs offices.

MEXICO, BAJA CALIFORNIA

ALFONSINA'S *'s (Senterfitt No. 147)* A packed dirt strip (with a few soft spots) right beside the beach which is completely isolated except for some 35 trailers and small cabins maintained by American fliers and a small restaurant run by Alfonsina Urquidez V. who leases the land from the Mexican government and subleases it to the Norte Americanos. Rustic accommodations are available. Gas is usually available and the food at the restaurant is safe. There are miles of lonely beaches that are excellent for camping (no facilities). The water is delightfully warm in summer and fall. A large slough next to the airstrip covers several square miles of interesting tidelands where thousands and thousands of gallons of sea water flow in and out of a small inlet with every tide. It is a very interesting area for birds, fish and other marine life. The tide covers the runway several times a month which keeps it hard packed. Frequent visitors use a tide calendar (52) to avoid arriving during the periods of temporary inundation. See also page 23 about salt water problems.

It is at the end of a very long and very rough road. There are two other small fish camps about 6 miles away but otherwise there is no trace of civilization for 50 miles. You can go a short ways down the beach and be completely alone. We have had a small cabin there for a number of years. We can taxi right up to the back door and the beach on the bay is just outside the front door. It has made an excellent get-away place for the whole family.

HAMILTON RANCH *(Senterfitt No. 90)* See page 61.

MELING RANCH *(Senterfitt No. 86)* Tops on our list of places to fly to! Included here because while it is not a camping place or even wilderness it is very isolated and gives one a very realistic feeling of having moved back some 100 years in time into the atmosphere of an old-time cattle ranch. Aida Meling is a very relaxed and gracious hostess. Excellent farm type food is served family style at long tables and rooms are very comfortable. There is usually a very interesting and compatible group of guests. (The last time that we were there our table companions included a Nobel Prize winner and a topless Go-go dancer!) The strip has a slightly rough approach but it is very long. A go-around could be difficult for an overloaded plane and watch out for irregular turbulence when landing in windy conditions. Reservations are usually necessary. You can write to the Meling Ranch at P. O. Box 224, Ensenada, Baja Cfa, Mexico but allow a lot of time. If you know any ham radio operators perhaps they can contact the ranch in the evening between 8 and 9 P.M. (Pacific Time) on 5 meter FM. Aida's call sign is XE2BY. Many ham operators in California talk to her regularly and can perhaps arrange a patch for you. (50)

TURTLE BAY (BAHIA TORTUGAS) *(Senterfitt No. 125)* An unexpected picturesque little fishing town in the middle of a great sandy wasteland. Again, not a wilderness area, but a good example of the unusual that you can visit easily only by private plane. Small relatively neat houses laid out in mini-city fashion with a fish factory, pier and boats. No food or accommodations. Strip wide, long and good (is used for large cargo planes for the fish factory output).

SCAMMON'S LAGOON *(Senterfitt No. 121)* A completely isolated strip adjacent to the famous winter whale region where many California grey whales come to breed and spawn. There is not much else to see or do but it is quiet and isolated in typical treeless desert country. From December through March the whales can be seen spouting and surfacing in the distance. You will need a larger boat than will go in a plane for close up views, but from the air the whales can be easily seen. Much under 1000 feet your plane will make them submerge, but above this altitude with a telephoto lens you can get good pictures. (See page 134.) The strip is dirt with a fair surface and rather short. Good approaches. The paved highway down the peninsula runs near but not actually to the strip. Non-campers can find good food and accommodations adjacent to the excellent paved strip at *Guerro Negro (Senterfitt No. 120)* about 15 miles north; stay there and fly down to see the whales.

MEXICO – MAINLAND

COPPER CANYON, SONORA *(Barrance de Corbi on WAC Chart CH-23)* This is the most isolated airstrip I have ever landed at. It is a long dirt strip on the top of one of the islands of land in the center of Mexico's "Grand Canyon" which is larger but less colorful than the one in the U. S. An extremely picturesque and beautiful spot. Good camping (no facilities) with beautiful trails of all degrees of severity. You will probably see a few of the Tarahumara Indians on the mountain trails. These are a hardy race that is currently being studied by teams of U. S. Physiologists because of their ability to run rapidly for many miles while subsisting on about 1500 calories a day of poor quality food [45]. About a mile away someone is intermittently working at building a little resort but the rudimentary roads make progress on it sporadic.

The strip seems longer than the 4400 feet indicated on the chart. It is quite wide and the surface is relatively smooth firm dirt. There are no obstructions rising vertically on approach, but the vertically falling terrain does constitute a definite hazard. It is thousands of feet nearly straight down on all sides with nothing resembling smoothness at the bottom. There are several minutes on approach and departure when you are completely dependent on your engine. I do not like to fly this way, but if any spot is worth this risk, this one is.

Caution: This is real mountain country. Don't attempt it if the weather is not perfect and nearly without wind.

If you camp be sure and have extra time available because you won't want to leave if there are any winds or other bad weather. Prolonged unfavorable weather is rare in this country but you could be delayed a day or two, and it's not a place to take *any* chances at!

EL FUERTE, SONORA *(WAC Chart CH-23; ONC H-22)* This is a sleeper. Most visitors go to nearby Alamos which is very interesting but is on the typical tourist circuit and overrun with Americans, although it does have very good accommodations. El Fuerte was the military post that protected Alamos when that was a thriving mining town. El Fuerte has an excellent paved strip, an adequate motel and the appearance and atmosphere of rural Spain without the crowds of tourists. Fishing on the nearby lake is reported to be very good.

GUANAJUATO, GTO. *(ONC J-24)* I have not personally visited here but several friends whose judgement I have found I can rely on have made trips recently and highly recommend it as a smaller Mexican city where there are very good accommodations and yet one can still get the true Mexican "atmosphere" away from the jaded tourist circuit. The strip is high and relatively short but the weather is apparently usually quite cool and approaches are reported good.

SAN JUANITO, CHIHUAHUA *(WAC CH-23)* A sleepy little Mexican mountain town that I always enjoy wandering through. Seldom does a tourist visit here, although the famous scenic Topolobampo – Chihuahua railroad makes a brief stop here.[46] Several small stores and a lumber-ranching atmosphere. The very good strip is nearer to town than the WAC chart indicates.[47]

SAN BLAS, NIARIT *(ONC-J-24)* One of the first really pictureseque little fishing towns that you reach after traversing the "great bald spot," as someone has described the northern part of mainland Mexico – (someone who has obviously only driven there and has not seen the beautiful little mountain spots that are available up there to the flier). San Blas has two ancient hotels with lots of charm and "atmosphere." There is also an excellent more modern motel right on the water with very good food as well as a rambly older hotel a few miles out of town. Ruins of an old fort are interesting to wander through and swarms of very colorful birds abound. Excellent surfing a few miles south, if you have wheels and can take a surf board in your plane; although I suspect that a surfer could manage to borrow or rent one from some of the buffs in town. This is a quite picturesque unspoiled little town with true Mexican feel which the lightplane flier can easily take advantage of.

CANADA

BELLA COOLA, BRITISH COLUMBIA *(WAC Chart E-15) Air Facilities Map of British Columbia*[48] The spectacular scenery here is the best that I have ever seen! It looks like Switzerland with an ocean added. Abrupt high mountains border a narrow green valley with little farms and homes beside a beautiful river. This little Norwegian settlement on a fiord north of Vancouver Island is at the end of a very rough 250 mile road from the Cariboo Highway. It is not true wilderness itself, but the little settlement is completely surrounded by wilderness. Fishing is apparently excellent and the harbour is full of boats of various kinds. The paved airport is excellent with good approaches. The safest route is to follow the road from Williams Lake but even so, the enroute terrain as you cross Tweedsmuir National Park is extremely mountainous. Some intrepid souls approach it over the water coming up the coast from Vancouver Island. Without floats or at least three engines this is not for me! Weather could be a problem, at times high fog and low clouds hang halfway down the mountains and close the passes for days.

BRAEBURN, Y. T. *(WAC Chart D-12) Yukon Air Facilities Map* (49) Here is an easy to find and easy to use little strip that combines nearly true wilderness and comfortable accommodations! At the south end of the long wide gravel strip Braeburn Lodge, known throughout this area for its fine food, also has rooms. Camping at the other (north) end of the strip is completely out of sight and sound of the lodge and completely isolated except for the Klondike Highway which runs a half mile away and is not a problem. Lakes with excellent fishing and woods to hike in are all around. An easy overnight stop for either camping or accommodations if you are en route between Whitehorse and Dawson City.

CLINTON CREEK (COUSINS STRIP), Y. T. *(WAC Chart D-12) (YAF Map* (49) *)* Also known as Mile 924 of the Alaska Highway. (Not to be confused with the Clinton Creek strip north of Dawson City.) This completely isolated strip about 8 miles from Whitehorse Airport is an excellent alternate for those who want to camp near Whitehorse and avoid the noise of the large airport. No camping facilities but it is peaceful and quiet. The immediate area is quite dusty but there are wooded and grassy areas nearby where you can avoid the dust (see photo page 66). Not much to do here other than walk in the woods and down to the nearby river and think of Robert Service who wrote many of his poems about this area. If you have wheels, you can reach a grocery store about five miles south on the Alaska highway. I pulled in there on my little Mo-ped with California plates and raised a few curious eyebrows when I took off with my purchases to join the northbound traffic which consisted of heavy duty rigs with extra spare tires, etc.

DOG CREEK, B. C. *(WAC Chart E-16) (AFM of BC No. 98)* (48) To me, a very relaxing and interesting spot. Site of an old RCAF early warning system airbase. The large old triangular field has been left nearly completely unchanged. The surface is gravelly with soft and occasionally muddy spots. Many buildings are still there with traces of the old orange and white World War II RCAF paint, and other momentoes of its active wartime life can be found lying about. You can taxi up an old road for ¼ mile and park your plane near some pine trees beside one of the old buildings and have your own private airbase! One very occasionally used dirt road skirts the far distant permimeter of the base and there is one farm about 5 miles away, not visible on the ground. Except for this you are entirely alone except for grazing cattle and a herd of semi-wild horses that will visit you with cautious curiosity. You can sit beside or in the old house and let your imagination see the RCAF bombers landing and taking off on training and patrol missions. The terrain is flat rolling country that feels open and free. You can wander *entirely alone* across the fields and through pine woods. Williams Lake is about 50 miles north and you can get its aeradio station nearly down to ground level.

"Accommodations" at Dog Creek. Probably about minus three stars by Michelin standards, but very relaxed and restful if you are camping.

INGENIKA RIVER, B. C. *(Not shown on current WAC Charts but is located on WAC CD-12 at spot where the Ingenika River reaches the wide Finlay River, just below the "29°E" figures. Shown as No. 165 on AFM of BC* [48]*.)* An excellent strip with easy approaches about in the middle of "the Trench" – a 250 mile cut off of the Alaska Highway between Prince George and Watson Lake which saves about 200 miles by avoiding cutting back to follow the highway back to Fort Nelson and Dawson Creek. A good part of it is over Williston Lake, one of the largest man made lakes which is an ecologist's nightmare that has covered thousands of acres of beautiful timberland and indian villages with water. Timber was not cut away ahead of the rising water and the entire large lake is nearly full of floating dead trees. Over most of the Trench logging roads or logging airstrips provide emergency landing facilities (a list and description is usually available at either Watson Lake or Prince George aeradio station offices). Repeater stations enable radio reception over a good part of it. The "follow the highway" advocates insist that it is safer to follow the highway. You will have to decide for yourself. I first took it with some hesitancy when it had good weather and the highway area did not. Personally I think it is wider and safer in many ways than the highway area. It makes you wonder why the highway was not run up this relatively flat valley instead of its circuitous route away from it but there were probably engineering or political reasons that are not apparent from the air. For either route be sure that the weather is good. It is rare that there are not some cumulus clouds and occasional rain showers, but the visibility should be good before you try it, with a ceiling of at least 5500' in order to maintain enough elevation to glide to a flat spot in case of emergency. Ingenika River has a considerable tribe of Indians and a trading post here. Backwoods country hospitality is apparent here and often a cabin and food is offered. There is often no charge but some payment should be made because the trading business is not flourishing at this time.

FORT WARE, B. C. *(WAC CD-12 shows its location as a water aerodrome only). (It is No. 131 on the AFM of BC)*[48] I thought a long time before I decided to include this one because it is a little known spot that is a wonderful combination of wilderness with interesting things and people, and it would be a shame for it to become really popular. It is on up the Trench another fifty miles or so, beyond where the lake stops and the river carries on. A small band of Indians was cut off by the rising waters of the lake and formed a village here much more separated from civilization than most Indian

On final at Fort Ware. What you see in this picture are just about the only signs of civilization for a long, long way in any direction.

villages. There are no roads for miles and the only approach was by foot or river boat until a dirt strip was built in 1974. Now you can land on a rough but usable strip and observe the customs of the Indians in about the most unspoiled setting that it is possible to find today. The Indians are friendly and curious. You will be as much an interest to them as they are to you. Canadian hospitality is the rule here, too, and you can relax and enjoy seeing a different world. If you can get one of the Indians, or perhaps some of the Van Somers family that run the small trading post here, to take you up the river in a boat there are some 25 miles of navigable waters through some of the most beautiful and isolated country you will ever experience. (See photos on pages 181 and 216.) Moose, bear and sometimes wolves seem to be abundant in this country.

PINE LAKE, Y. T. *(WAC Chart D-12) (AMF of BC No. 279)* (48) I flew over this on several trips and looked at it as an emergency landing spot, but was not impressed with it from the air as a potentially pleasant camping spot. One night, darkness and bad weather ahead made me stop there and I found that, on the ground, it was a beautiful and fascinating place. It was a supply strip for building the Alcan Highway in 1943 and it has been maintained for summer emergency use. A long wide gravel strip with adequate approaches right next to the wide slow Rancheria River with beautiful camping and fishing spots right near the aircraft parking area. Nearby on the other side is the smaller Swift River which runs to the Yukon River and eventually the Bering Sea. The Rancheria river, here only a half mile from the Swift, runs into the MacKenzie River and ends up a thousand miles away in the Beaufort Sea.

The surrounding mountains are moderately high, forested and lonely. If you have wheels, a road runs back some 18 miles to an area of abandoned mining activity past some beautiful isolated lakes. On the ground the Alaska highway is not apparent at all, being some miles away through a dense forest. Fresh bear tracks appeared one day in a muddy area beside the plane between breakfast and lunch while we were away from it exploring. (See photo page 137.) For those who like to fish, wander through nearly virgin forest or watch isolated mountains and lakes, this could be a place to spend a long time. No particular flying problems or idiosyncracies about this strip. This is really a good example of isolation and quiet wilderness.

ALASKA

EAGLE VILLAGE *(WAC Chart CC-9) (Dawson Sectional)* (Strip is in the center of town and should not be confused with the better strip just to the south called Eagle.) Not true wilderness but a very picturesque outpost of civilization on the Yukon River as it flows into Alaska from the Yukon Territory. You can get customs clearance here at the post office into the U. S. from Canada. Land at the strip (which is downhill to the river) in the center of the village. Has the flavor of old time Alaska with very few of the encroachments of modern "civilization."

GLACIER PARK *(WAC Chart CD-11) (Anchorage Sectionals* do not always show it. Located just west of the *Sheep Mountain NDB.)* A relatively poor rocky strip with poor parking areas, about the minimum that I like to use. About one mile from the receeding edge of the Matanuska Glacier which is worth the hike. (See photo page 144.) You can see the conditions as the glacier melts and leaves its rocks and glacial silt (which can be seen blowing off in the wind while still wet)! Climbing up on the slippery glacier you can crawl down in crevasses and cracks in the ice. Note how the dark rocks and gravel on the surface of the glacier have absorbed more heat than the white ice surface and melted down into grooves and holes. A rather weather beaten sign at the strip says, "$10.00 per landing," but there was no one around to pay. This attitude plus the difficulty in getting off the strip to park makes it a poor bet for camping but exploring the glacier is well worth the stop. If you can arrange satisfactory parking, there is a public camping ground nearby.

GOLD BENCH *(WAC Chart CC-9) (Fairbanks Sectional)* A dirt strip of fair quality at the site of an extensive but abandoned goldmine. True isolated wilderness on a picturesque river in beautiful country. Old mining buildings in fair condition that show evidence of damage from large bears. Use caution if you wander around in the surrounding underbrush! This is a good example of the type of strip of which there are many in the mining areas of Alaska. Many gold mines used lightplanes to carry in and out supplies and personnel, and many abandoned mines still have these strips in poor to fair condition which make excellent isolated camping spots. Most of them are associated with rivers and mountains. There is very little, if any, maintenance on most of them, so use caution. (See photo on page 111.)

KOBUK *(WAC Chart CC-9) (Fairbanks Sectional)* Most of the villages you will visit in the north country will be inhabited by Indians. For contrast, here is one that is one of

On high final for the airstrip adjacent to Kobuk. This strip is well maintained and obviously frequently used. (Contrast this to the strip at Gold Bench on page 111), which apparently receives no maintenance and just keeps itself adequate enough for the occasional camper or hunter who may land there.)

the better and more interesting Eskimo villages. On the flat, broad Kobuk River, this town's main occupation is catching and smoking salmon. A sign on the schoolhouse says, "Dog teams and skimobiles keep out of schoolyard." The people are relaxed and friendly and the whole atmosphere helps you realize that you are north of the Arctic Circle. No Accomodations. You could probably camp here but you might be the object of a lot of attention. Like many other strips in the interior of Alaska, this one can be reached only after flying over a lot of completely uninhabited country. I felt that there were always adequate emergency landing spots for me, but you would have to look it over as you proceed and decide for yourself. (See page 51.)

LAKE LOUISE *(WAC Chart CD-11) (Anchorage Sectional)* Very isolated country. Rather flat country for Alaska but on the side of a beautiful lake at the end of a side road from the Glenn Highway, One functioning lakeside lodge and marina about two miles from the strip. Another abandoned lodge nearer with fascinating old buildings and cabins. A beautiful lake in quiet country. A sign at the strip says, "No Camping," but no one is around for miles to care. You can find excellent lakeside camping spots (without facilities) within a half mile of the strip. Gulkana FSS has diagrams of this strip (as well as others nearby) and you can receive Gulkana Radio nearly to the ground at Lake Louise.

McKINLEY PARK *(WAC Chart CD-11) (Anchorage Sectional)* (Not to be confused with *Mt. McKinley Park Strip,* a private strip a few miles south.) This one can be a little hard to find because turbulence in the well named *Windy Pass* can be distracting and there are several strips in the area that look similar to this one. Be sure and note the definite road running west into the park and also note the prominent railroad cars permanently moored at the hotel just a little to the west of the strip. The railroad station is also prominent. This is a good airstrip beside the highway and railroad at the entrance to McKinley National Park. It is therefore not true isolated wilderness but it does have many worthwhile things to see. McKinley has by far the fewest visitors of any of the major National Parks and thus has the maximum of park advantages and the minimum of the disadvantages.

You can camp in relative quiet fairly near your plane, if you park it near the south end. Auto traffic is, thankfully, restricted in the park. In season you can take a twice daily animal tour bus for a moderate fee which includes a catered meal along the route. This is usually full with the retired set who are in abundance around the park hotel. Or, you can make up your own lunch and join the backpacking crowd and take the free park service bus. You can get off at any point and look around and jump on another bus an hour or so later. The guides are young college students and interested in the park and its animals and will stop whenever wild animals are seen. This, like fishing, will vary. Some days many interesting species are seen, other days are slim. Some of the tent-camping-only campgrounds look interesting if you can solve the logistics problem of your distant plane. There are the other National Park type activities which are here centered primarily around wild animals and their activities. The McKinley sled dogs seem to be an almost separate breed and are a must see for any dog freaks. (See photograph page 222.)

NINILCHIK *(WAC Chart CD-11) (Seward Sectional)* A good gravel strip on a gravel road about 4 miles from Ninilchik Village on the coast of the Kenai Penninsula. The village has a quaint Russian church and a small fish cannery. Good opals are said to be occasionally found on the nearby beach. The strip is isolated except for small homestead type farms in the surrounding woods. Good for camping, but no facilities.

RAMPART *(WAC Chart CC-9) (Fairbanks Sectional)* An isolated fishing village on the shores of the Yukon a relatively short flight from Fairbanks. (Rivers in Alaska tend to be broad, not for excess water capacity from spring runoffs as in other places, but because the frozen tundra keeps the river shallow.) Indians live in permanent log cabins on the river shore. They raise their own somewhat different breed of sled dog and do not want visiting male dogs to enter the village. There is a fish wheel or two on the river which is worth seeing if you are not familiar with them. Use care in the brush covered riverbanks; there are many signs of bears. Beautiful country on the dirt road to the south with an occasional little gold mine.

THE BAHAMAS

NORMAN CAY *(WAC Chart CH-25)* This is the other airport listed here which I have not personally visited, but I included it because of what I felt were very reliable reports that it was an excellent Caribbean fly-in resort of luxurious (and expensive) calibre with good food and accommodations and friendly service.

SANDY POINT *(WAC Chart CH-25)* This appeared to me to be a good camping strip on an excellent beach at an isolated point on Great Abaco Island with no one anywhere near it. Many large and colorful seashells on the beach.

STANIEL CAY *(WAC Chart CH-25)* An excellent strip which seems to take up most of the island. Two small resorts cater to fliers and yachtsmen. We stayed at the Staniel Cay Yacht Club Marina which had very good accommodations and food. There is an interesting colony of friendly black folks many of whom have spent nearly all their lives on this tiny island. A couple of tiny grocery stores and a post office. The water is warm and skin-diving excellent. The movie "Thunderball" was taken on a nearby island which has a large interesting cave for diving. Camping is not allowed. (See photo on page 229.)

FOOTNOTES

(1) Quotation by David Brower from *BAJA CALIFORNIA AND THE GEOGRAPHY OF HOPE,* by Joseph Weed Krutch and Eliot Porter, copyright@1967 by the Sierra Club (used with permission of the publisher).

(2) *National Wildlife Federation Magazine,* September 1976

(3) *Los Angeles Times.* Front page, November 12, 1976

(4) See pages 23 and 234

(5) A few of many are:

Montana Airport Directory. Published annually by the Montana Aeronautics Commission, P. O. Box 1698, Helena Montana 59601. A small very complete looseleaf directory with airport diagrams and information for nearly every strip in Montana. An excellent Montana Aeronautical Chart supplements the book.

Idaho Airport Facilities published by the Idaho State Department of Aeronautics, 3013 Airport Way, Boise, Idaho 83705. Airport diagrams and much good information on flying in Idaho. Well worth the small charge for it. The *Idaho Aeronautical Chart* supplements this and gives a very useful list of Air Marked Forestry Lookout Towers which is more extensive than the listings on the regular Sectionals.

(6) *Survival Manuals.* Each branch of the armed forces seems to have its own. I like AFM-64-5 (D 301.7:) available for $2.50 from Superintendent of Documents, U. S. Government Printing Office, Washington, D. C. 20402. Also good is *Down But Not Out* obtainable from Information Canada, Publishing Division, Ottawa, KIA 0S9. Ask for Catalog No. DC-3-2270 and include check for $3.00 to Receiver General for Canada.

(7) 5. . . . an aircraft when flown in a sparsely settled area shall carry

(a) for each person carried, five pounds of concentrated food or its equivalent that is of high nutritive value and not subject to damage by heat or cold and has been inspected by the owner of the aircraft or his representative not more than six months prior to the flight, the food to be contained in a waterproof package bearing a tag or label upon which is a certification of such inspection;

(b) adequate cooking utensils and mess tins;

(c) matches in a waterproof container;

(d) portable compass;

(e) an axe of two and one-half pounds or heavier with a twenty-eight inch handle;

(f) thirty feet of snare wire;

(g) a sharp jack knife or hunting knife of good quality;

(h) additional equipment during summer conditions consisting of
(i) four trawls, two fishing lines with an assortment of hooks and sinkers and a fish net of not more than two inch mesh, and
(ii) sufficient mosquito nets to accommodate all persons carried;

and additional equipment during winter conditions.

(8) Alaska law requires the following minimum items for cross-country flying during the summer months:

1. food for each occupant sufficient to sustain life for two weeks;
2. one axe or hatchet;
3. one First Aid Kit;

4. one pistol, revolver, shotgun, or rifle, and ammunition for same;
5. one small fish net and an assortment of tackle such as hooks, flies, lines, sinkers, etc.;
6. one knife;
7. two small boxes of matches in waterproof container;
8. one small mosquito headnet for each occupant of the plane;
9. two small signalling devices, such as colored smoke bombs, railroad fuses, or Very pistol shells, in sealed metal containers;
10. flashlight with spare bulbs and batteries.

The following additional minimum items are required from October 15 to April 1:

1. one pair of snowshoes;
2. one sleeping bag;
3. one wool blanket for each occupant over four years of age.

(9) *"Trail Shops"* in your locality are best found by looking in the Yellow Pages under "Camping Equipment." Catalogs for mail order companies are offered in most any outdoor magazine. Some old standbys are: *Eddie Bauer,* P. O. Box 3700, Seattle, Washington 98124; *Holubar Mountaineering,* Box 7, Boulder, Colorado 80306; *Kelty,* 1801 Victory Blvd., Glendale, California 91291; *L. L. Bean,* Freeport, Maine has interesting and unusual items but the catalog is disorganized and not indexed. *REI, Inc.,* 1525 11th Avenue, Seattle, Washington 98122. (See page 122.)

If you order by mail always allow plenty of time before you really need the things; none of these seem to be very efficient in getting orders out very fast, and especially when you tell them you are in a hurry it always seems to take longer.

(10) Water purification. The simplest way is to strain it (if necessary) and boil it for at least 12 minutes (longer at higher altitudes). There are several types of "water purification" tablets that can be obtained from most drug stores. Use the directions and give the tablet plenty of time to work. Some travelers carry a small dropper bottle of ordinary liquid laundry bleach. This can be taken into restaurants and dropped right into the glass of water on the table. Use 3 drops for a glass (and 10 drops for a quart). It takes some time for it to work (20 minutes or so), so allow for this. (Keep the bottle well wrapped so it won't leak on other gear.) The whole subject, as discussed on page 153, is really a medical uncertainty, but these are the popular and accepted methods.

(11) *Dredging for Gold* "The Gold Divers' Handbook," by Matt Thornton. Published by Keene Industries, 9330 Corbin Avenue, Northridge, California 91324. A complete and very readable guide to finding gold with pan, dredge or metal detector.

(12) *REI, Inc.,* 1525 11th Ave., Seattle, Washington 98122

(13) Pine Lake (See page 240)

(14) Various types of fishing are done at the following places:

Capistrano (page 213); Cavanaugh Bay (page 232); Alfonsinas (page 234); El Fuerte (page 236); San Blas (page 236); Bella Coola (page 236); Braeburn (page 237); Fort Ware (page 239); Pine Lake (page 240); Kobuk (241); Lake Louise (242); Rampart (page 243); Normans Cay (page 243); Sandy Point (page 243); Staniel Cay (243).

(15) *The Wolf* by L. David Mech. Published for the American Museum of Natural History, The Natural History Press, Garden City, N. Y., Library of Congress Card No. 73–100043, ISBN 0–385–08660–1. A scientific yet very readable book fully covering the wolf. (Reading this will help you understand your dog.)

(16) *National Geographic Magazine,* Vol 148, No 3, September 1975, page 428, "Alaska's Big Brown Bears" by Allan Edbert and Michael Luque.

(17) *A Field Guide to Animal Tracks* by Olaus J. Murie. Houghton Mifflin Co., Boston, Mass., ISBN 0–395–19978–6. A paperback guide very helpful for those who want to determine what animal made those tracks along with some basic information about the animal.

(18) *Pacific Search Magazine* Vol 11, No 2, November 1976, page 49, item by Terry Johnson

(19) *Remington Sporting Firearms & Ammunition Catalog.* Yearly by Remington Arms Co., Inc., Bridgeport, Conn. 06602

(20) *American Zoologist* Vol 7, No 2, May 1967. 16 papers by various authorities on wolves at a symposium given in August 1966 by the Animal Behaviour Society, the Ecological Society of American and the American Society of Zoologists. The papers are quite technical, but very complete.

(21) See your veterinarian or write *United Pharmacal Co.,* 306 Cherokee, St. Joseph, Missouri 64504 OR 83366 La Mesa Blvd., La Mesa, California 92041 for a catalog of animal medicine.

(22) *Merck Veterinary Manual,* Merck & Co., Rahway, New Jersey

(23) *Merck Manual.* Obtainable through your bookstore or from Merck & Co., Rahway, New Jersey.

(24) Most larger commercial first aid kits are adequate but many contain items that are unnecessary or obsolete, such as smelling salts and iodine or merthiolate for cuts, which add useless weight. Be sure and have plenty of sterile gauze pads of several sizes and sterile roller bandages. Bandaids are always used. A triangular cloth for an arm sling is often included, but one could be made up out of other things if needed. An elastic roller bandage is often useful. Adhesive tape is essential. 70% alcohol and a needle for splinters are important. There will probably be more complex items but these will depend upon your medical talents.

(25) *The New England Journal of Medicine,* Vol 294, No 24, June 10, 1976, page 1340. One of the oldest and most respected medical journals gives a summary of the latest in Travelers' Diarrhea. It may be too technical for many non-medical readers but it might be a good one to suggest to your doctor if he is uncertain on the latest aspects of this subject.

(26) Ibid page 1299

(27) See *REI, Inc. 1976 Catalog*

(28) *AOPA Pilot* "September 9, 1963, page 90, "Scuba Diving and Flying"

(29) *AOPA Airports USA,* 7315 Wisconsin Ave., Washington, D. C. 20014

(30) *A Short History of Mexico,* J. P. McHenry, Dolphin Books, Doubleday, Garden City, N. Y. 1962; Library of Congress Card No. 62–10467. A relatively objective history of Mexico that reads like a novel. Small paperback size good for plane travel.

(31) A credit card sized plastic card which provides insurance coverage by Tepeyac, a large Mexican Insurance Company, for your plane (or other vehicle) if you phone in prior to leaving the United States to activate the coverage. You can get coverage for a specified

number of days or leave it open until you call again to close it. Later you are billed for the time covered. If you have an accident in Mexico the Tepeyac adjusters take care of it. It is very popular because it is so convenient and it apparently works well, although you wonder if, especially in smaller places, a definite written policy wouldn't be accepted easier. Mexicard is available from MacAfee and Edwards, 3105 Wilshire Blvd., Los Angeles, California 90005, (213) 388-9674.

(32) *Airports of Baja California* by Arnold Senterfitt. Equivalent in size and format to Jepp charts but are for VFR and tourist and other information is included. Update service available.

Airports of Mexico is similar, but covers the mainland of Mexico. At many FBO's or from *The Pathfinders,* P. O. Box 967, Lakeside, California 92040.

(33) ONC charts can be ordered from your FBO or your usual source for the regular sectionals.

(34) *The Bantam New College Spanish & English Dictionary.* Bantam Books, Inc., 666 Fifth Ave., New York, New York 10019. A thick pocket paperback. A little harder to carry than some of the very tiny ones, but more complete and adequate in vocabulary.

(35) Canadian Aircraft Owners and Pilots Association, Box 734, Ottawa, Ontario, Canada, KIP 5S4.

(36) Canadian VFR Chart Supplement (GPH 200A). Single copy $1.25 from Canada Map Office, 615 Booth Street, Ottawa Ontario, KIA OE9. (Order well in advance.)

(37) The Canada Map Office, 615 Booth Street, Ottawa, Ontario, KIA OE9

(38) Air Tourist Information from Ministry of Transport, S. L. P. P., Ottawa, Ontario, KIA ON8

(39) See pages 252 and 254.

(40) AOPA Flight Department, 7315 Wisconsin Avenue, Washington, D. C. 20014

(41) *The Milepost* published annually by the Alaska Northwest Publishing Company, 130 Second Avenue South, Edmonds, Washington 98020. A very complete mile-by-mile description of all major land routes in Alaska and northwest Canada. Slanted to the highway traveler, it does contain much information of interest to those visiting these areas by lightplane. The same company publishes numerous periodicals and books with beautiful photography and articles on this region.

(42) *Alaska and Western Canada Map* from any AAA office.

(43) Most flying periodicals have advertisements from various Florida FBO's who seem anxious to serve and to advise fliers going to The Bahamas. I have found unusually excellent service and help at Tilford Flying Service at Palm Beach International Airport. Jim Tilford who built it all up is retiring but the new owner, Pete Benoit, seems to be intent on maintaining their famous high standards.

(44) *Bahamas Air Navigation Chart* from nearest Bahamas Tourist Office or by mail from their office at 1730 Rhode Island Avenue, N. W., Washington, D. C. 20036; in Canada, 85 Richmond Street West, Toronto, Ontario M5H 2C9. Toll free phone, 800/327-0787.

(45) *The Physician and Sportsmedicine,* Vol 4, No 2, February 1976, page 38, "Indians Who Run 100 Miles on 1,500 Calories.

FOOTNOTES

(46) *Destination Topolobampo* by John Leeds Kerr. Published in 1968 by Golden West Books, San Marino, California 91108, Library of Congress Card No. 68–29992. The history of the building of the *Kansas City, Mexico and Orient Railway,* of which the *Chihuahua al Pacifico* is a remnant. Good background reading for anyone planning to take this spectacular train.

(47) *Take The Train to Copper Canyon* by Don Downie, *AOPA Pilot,* April 1977, page 75. Describes a very spectacular train trip to the interesting country and natives of this area. We do not agree with the writer's pessimistic view on flying in this area. It is mountainous country that needs extra caution (see Copper Canyon on page 235), but with care you can fly up and get away from the crowds in and around the train and see Tarahumaras who are more interesting than the tourist-jaded ones around the train stops.

(48) *Air Facilities Map of British Columbia* (AFM of BC). Distributed free by the British Columbia Aviation Council, International Airport South, Vancouver, British Columbia, Canada. Shows the location and gives a good description of nearly every airstrip in British Columbia with a description of facilities. Use in conjunction with the WAC charts because terrain is not indicated. It also shows blue diamonds for "preferred VFR routes" (some of which are laid out by braver souls than I am!)

(49) *Yukon Airport Directory* from Yukon Department of Tourism and Information, P. O. Box 2703, Whitehorse, Yukon Territory. Y1A 2C6

(50) *Where the Old West Never Died* by Paul Sanford. The Naylor Company, San Antonio, Texas 1968. An illustrated history of the Meling Ranch.

(51) See page 22 for diagram.

(52) Times of tides are different in the Gulf of California. An attractive tide calendar is available each year for $2.50 from Tide Calendar, Printing-Reproductions, University of Arizona, Tucson, Arizona 85721.

ADDITIONAL REFERENCES

In addition to the references given in the footnotes, the following will give added information to help plan your trip and make it more enjoyable.

LOWER UNITED STATES

Western States Fly-In Campground Directory. 1974 by Fly 'N Camp Enterprises, P. O. Box 6055, Burbank, California 91510. A very simply done saddle stitched little paperback (most of the text is reproduced handwriting) that does have a lot of good information about specific airports with nearby camping facilities in ten western states. It covers primarily strips with established campgrounds which are often located some miles from the strip.

California State Park System Guide from Department of Parks and Recreation, P. O. Box 2390, Sacramento, California 95811 has information on California State Campgrounds, some of which are easily accessible by lightplane.

MEXICO

Baja by Air by Allen and Phyllis Ellis, by Pan American Navigation, 12021 Ventura Blvd., North Hollywood, California 91604. One of the pioneer flying guides to Baja published in 1967. It has been somewhat outdated by the many changes and developments in Baja since, but it still contains useful background information and philosophy for Baja flying.

ADDITIONAL REFERENCES

The Sea of Cortez by Ray Cannon. Published by Lane Magazine and Book Company, Menlo Park, California. The classic book about Baja California by Sunset Magazine. Has only one page of brief private plane information but is recommended for anyone about to fly to Baja for the first time because its text and excellent pictures really tell what Baja is all about.

ALASKA

Plan-A-Flight to Alaska by George and Deloris Crowe. Published in 1968 by Plan-A-Flight Publications (mail returned from address given in book in 1977). Contains much helpful information and airport diagrams for major strips along the Alaska Highway. Does not cover any territory in Canada or Alaska not on the highway.

Supplement Alaska published every eight weeks as a supplement to the AIM Part I and to en route charts and sectional maps. A middle sized booklet giving diagrams and terminal information for most major airports as well as much other flying information for Alaska and western Canada. Obtainable at major FSS stations in Alaska or from National Ocean Survey, Route C—44, Riverdale, Maryland 20840.

THE CARIBBEAN

Caribbean Flight Guide by Randal Agostini. Caribbean Flight Guide Co., P. O. Box 191, Port of Spain, Trinidad, West Indies or by mail from Sporty's Pilot Shop, Batavia, Ohio 45103. A compact little looseleaf book with diagrams and information about many airstrips in the Caribbean. Helpful for those who like all this information collected in one spot with a diagram of the airport layouts. (If you order one, be sure that you get a *complete* one. I was sent one with less than half the listed contents, apparently an oversight during a recent changeover to the looseleaf format.)

GENERAL

AOPA Flight Department, 7315 Wisconsin Ave., Washington, D. C. 20014 has good booklets on *Alaska, Canada, Mexico, The Bahamas, West Indies* and *Latin America,* containing much practical and essential information about each area. If anything they are too complete and sometimes make flying to these areas seem much more complicated than it really is. Still, they do have much information that is needed and is hard to get from any other source.

Backpacking, One Step at a Time by Harvey Manning. Vantage Books, Random House, N. Y. ISBN 0—394—72033—4. Slanted toward the backpacker, this is a very readable yet complete book on all aspects of camping and camping equipment.

PROPPING CHECK LIST

1. Relax and slow down.
2. Before propping, try and find out why the battery is dead, so that the engine can hopefully charge it up again when it starts. If there is no obvious cause, such as a master switch left on, excess radio or light use on the ground, etc. clean the battery terminals and tighten them well. Inspect the wires to the generator or alternator. If any are loose or broken, try to repair them. If more than one is loose be sure replacements are in proper locations.
3. Load your plane and put passengers aboard. You don't want to stop the engine once it starts; and loading baggage and people with a rotating prop is dangerous.
4. Put the hand brake on hard and, if possible, have someone hold the foot brakes hard. This person should be shown how the brakes work, and also how to turn off the mag. switch if necessary to stop the engine suddenly. Do not depend on the brakes alone. *Be sure* that the plane is otherwise secured! Chocks in front of main gear wheels. (Not on nosewheel – too near prop.) A sturdy rope tied snugly to a solid object behind the plane is best.
5. If an external power source, or jumper cables and another battery is available, always use this at this point instead of propping. Be sure the polarity is not reversed and, again, watch out for the prop when the engine starts.
6. Open the throttle slightly more than normal to try and be sure the engine will keep running after it starts. If it stops (or if the engine is already warm) it may be too hard to prop it again. Be sure throttle is locked.
7. If the engine is carburetted, with all switches off and with the key in your pocket, make two hard quick pumps with the priming pump. Then slowly turn the prop in the proper direction, (the fatter side of the prop is the leading edge). Carefully, (an engine can possibly start even with the switch off), turn the prop one revolution for each cylinder. This puts a "charge" in each cylinder and makes starting easier; and you need it as easy to start as possible. Leave one blade at the 11 o'clock position.
8. Put key in ignition and turn it on. Keep master switch off. Stand in from the prop, and slightly to the side, in such a position that you have good solid footing and can turn the blade easily, and where all of you will be clear of the prop when it turns. Grasp the prop with both hands and pull it through from above you. Pull it quickly down (about a third of a complete rotation) letting go and pulling your hands away from it, at the same time backing away from it carefully.
9. If it starts, move safely away and keep away from it.
10. If it doesn't start, with the *key off,* and keeping all parts of your body away from its potential path, move the prop carefully back to a blade high position so you can start again.
11. Repeat steps 8, 9 and 10 as necessary.

APPROXIMATE FUEL CONVERSION

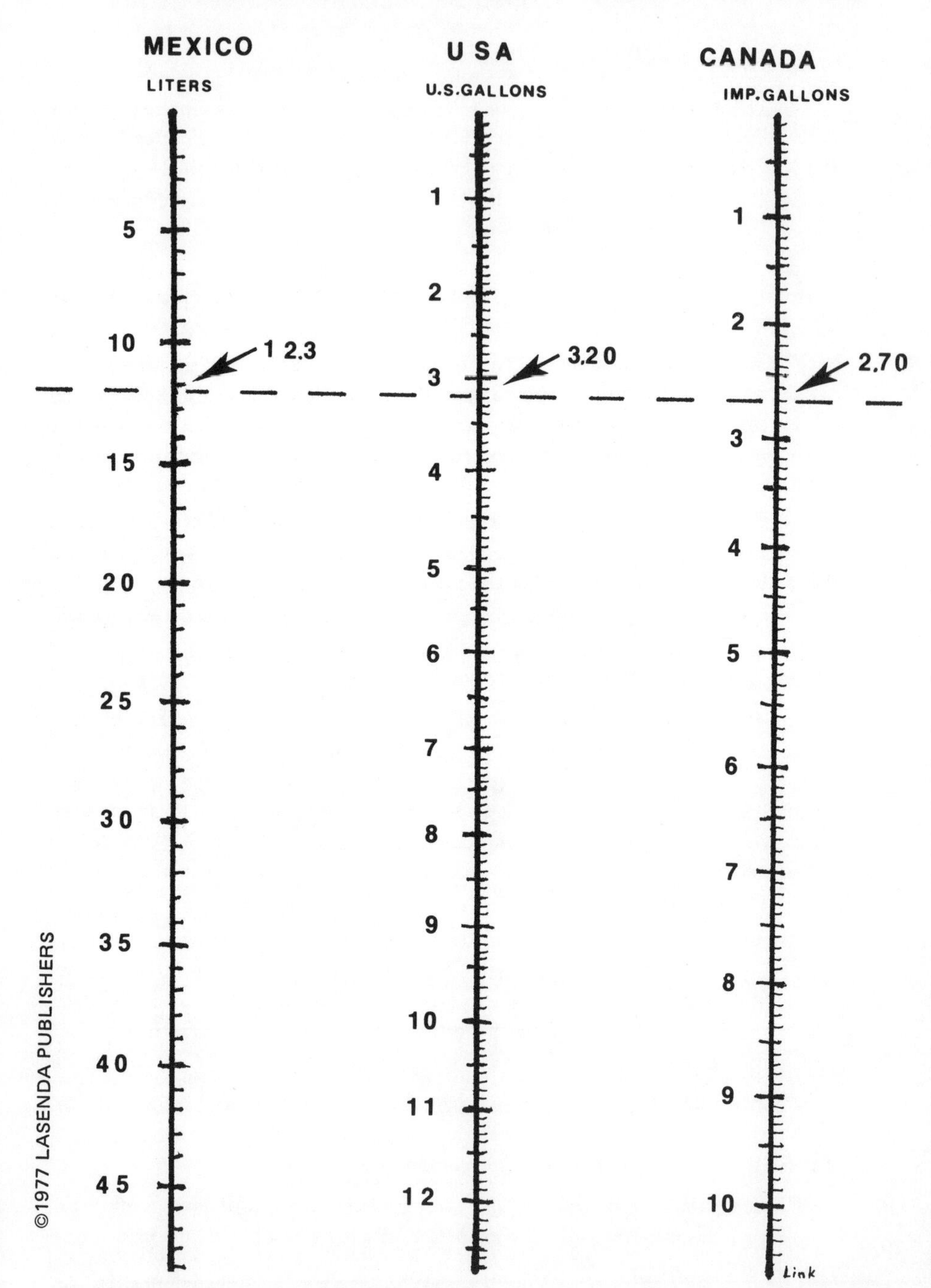

The illustrative line shows the conversion between any two of the three different measuring units. (Use like a slide rule – the line could indicate 3.2 gallons, 32 gallons or [you should be so lucky!] 320 gallons. The other units would, of course, correspond, i.e., 12.3, 123 or 1230 liters; or 2.7, 27 or 270 Imperial gallons.)

PROPPING CHECK LIST

1. Relax and slow down.

2. Before propping, try and find out why the battery is dead, so that the engine can hopefully charge it up again when it starts. If there is no obvious cause, such as a master switch left on, excess radio or light use on the ground, etc.), clean the battery terminals and tighten them well. Inspect the wires to the generator or alternator. If any are loose or broken, try to repair them. If more than one is loose be sure replacements are in proper locations.

3. Load your plane and put passengers aboard. You don't want to stop the engine once it starts; and loading baggage and people with a rotating prop is dangerous.

4. Put the hand brake on hard and, if possible, have someone hold the foot brakes hard. This person should be shown how the brakes work, and also how to turn off the mag. switch if necessary to stop the engine suddenly. Do not depend on the brakes alone. *Be sure* that the plane is otherwise secured! Chocks in front of main gear wheels. (Not on nosewheel – too near prop.) A sturdy rope tied snugly to a solid object behind the plane is best.

5. If an external power source, or jumper cables and another battery is available, always use this at this point instead of propping. Be sure the polarity is not reversed and, again, watch out for the prop when the engine starts.

6. Open the throttle slightly more than normal to try and be sure the engine will keep running after it starts. If it stops (or if the engine is already warm) it may be too hard to prop it again. Be sure throttle is locked.

7. If the engine is carburetted, with all switches off and with the key in your pocket, make two hard quick pumps with the priming pump. Then slowly turn the prop in the proper direction, (the fatter side of the prop is the leading edge). Carefully, (an engine can possibly start even with the switch off), turn the prop one revolution for each cylinder. This puts a "charge" in each cylinder and makes starting easier; and you need it as easy to start as possible. Leave one blade at the 11 o'clock position.

8. Put key in ignition and turn it on. Keep master switch off. Stand in from the prop, and slightly to the side, in such a position that you have good solid footing and can turn the blade easily, and where all of you will be clear of the prop when it turns. Grasp the prop with both hands and pull it through from above you. Pull it quickly down (about a third of a complete rotation) letting go and pulling your hands away from it, at the same time backing away from it carefully.

9. If it starts, move safely away and keep away from it.

10. If it doesn't start, with the *key off,* and keeping all parts of your body away from its potential path, move the prop carefully back to a blade high position so you can start again.

11. Repeat steps 8, 9 and 10 as necessary.

Cut on dotted line. Cover with an adhesive plastic laminating sheet and keep in plane for use when needed.

From: FLYING. . .OFF THE PAVEMENT by Link Grindle; Lasenda Publishers, 32331 Coast Highway, South Laguna, CA 92677

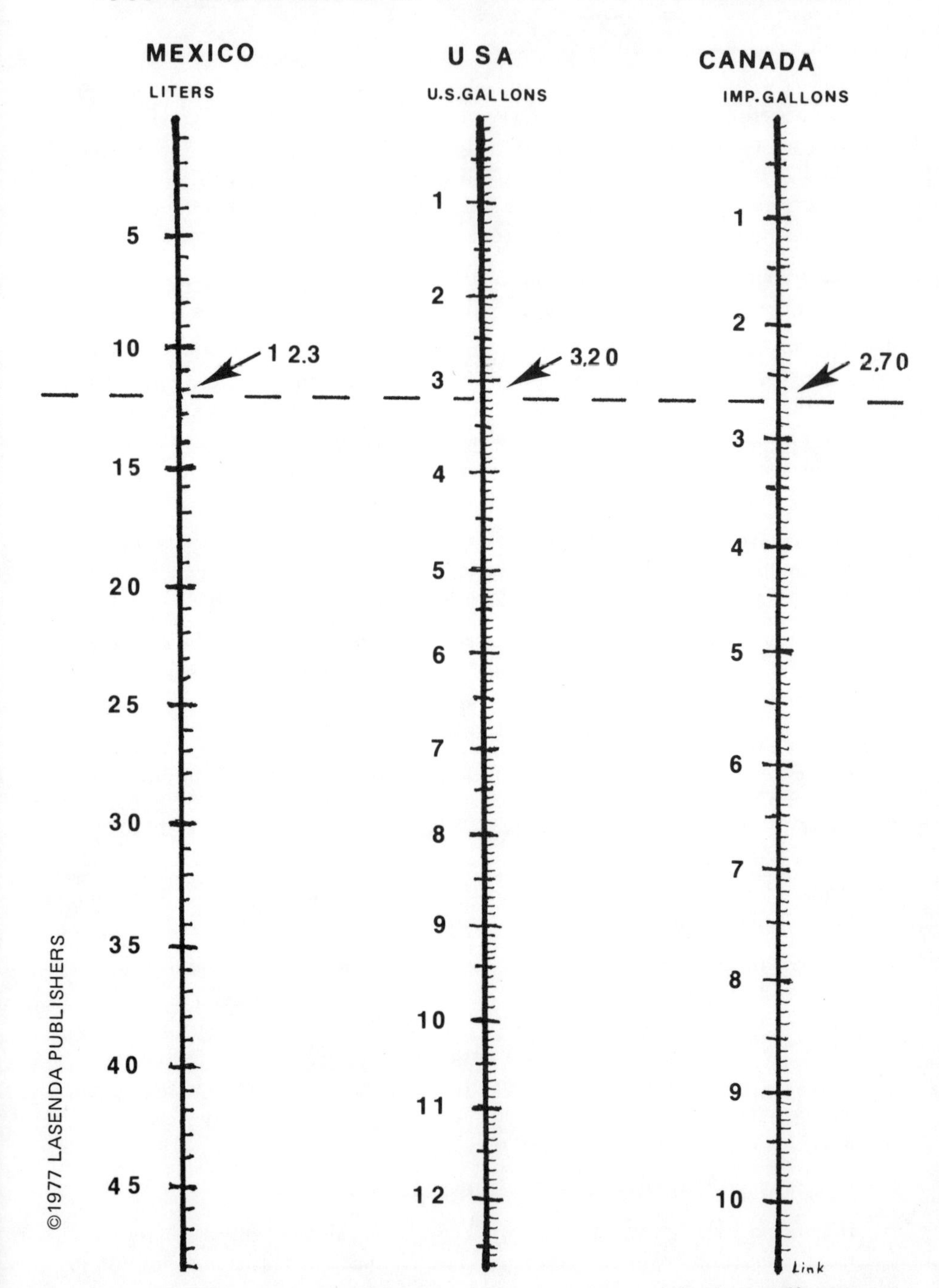

The illustrative line shows the conversion between any two of the three different measuring units. (Use like a slide rule – the line could indicate 3.2 gallons, 32 gallons or [you should be so lucky!] 320 gallons. The other units would, of course, correspond, i.e., 12.3, 123 or 1230 liters; or 2.7, 27 or 270 Imperial gallons.)

INDEX

INDEX

In-and-Out certificate, 220